Sustainable Development Indicators

Applied Ecology and Environmental Management

A SERIES

Series Editor
Sven E. Jørgensen
Copenhagen University, Denmark

For more information, please visit: www.crcpress.com/Environmental-Management-of-Marine-Ecosystems/Islam-Jorgensen/p/book/9781498767729

Sustainable Development Indicators
An Exergy-Based Approach

Søren Nors Nielsen

CRC Press
Taylor & Francis Group
Boca Raton London New York

CRC Press is an imprint of the
Taylor & Francis Group, an **informa** business

First edition published 2020
by CRC Press
6000 Broken Sound Parkway NW, Suite 300, Boca Raton, FL 33487-2742

and by CRC Press
2 Park Square, Milton Park, Abingdon, Oxon, OX14 4RN

© 2020 Taylor & Francis Group, LLC

CRC Press is an imprint of Taylor & Francis Group, LLC

ISBN 978-0-367-25735-4 (hbk)
ISBN 978-0-429-28958-3 (ebk)

Typeset in Palatino
by Apex CoVantage, LLC

Contents

Foreword

The book in your hands contains the scientific background—"philosophy" and methodological issues—to a project with the title "Environment and Sustainability Year 2012—New Development of Methods and Tools for Estimation of Sustainability" financed by the VELUX foundation in Denmark and carried out at the Energy Academy on the island of Samsø from February 2012 to June 2013. This project outlined a plan for analyzing the sustainability status of a relatively small society with fixed boundaries, in this case an island. We believed this would be an appropriate platform from which to obtain a proper overview of the problems arising during the development of such a method. The basic idea is to identify a measure by means of which the level of sustainability can be indicated, the indicandum being understood as the degree to which certain sustainability criteria—in this case, growth or steady state with respect to exergy—has been met. The result was a methodological framework ready for upscaling and testing for robustness. This is now being carried out while the methodology is employed as part of the estimations being carried out at the Mid-University in Östersund, Sweden to develop procedures for shaping a sustainable region of Jämtland.

An abbreviated acronym comes out as "Easy-NDOMATES" (*ndomates* [νδοματεσ] means tomatoes in Greek), and this is quite symbolic, since this part of the project report fulfils an important wish from the people living on Samsø, namely that the project would serve to identify the "lowest hanging fruits"—the initiatives that would be the easiest to implement and hence most sensible to start with in the process of achieving sustainability.

The metaphysics of the project is clearly rooted in a materialistic interpretation of sustainability. That is that the environmental pillar comes first, and all other forms of sustainability are secondary to this. Although social and economic sustainability initiatives will interact with environmental sustainability, they will in the end only be able to persist if the demands of environmental sustainability are fulfilled. As a consequence of working with such a view, a society is sustainable only if it is self-sustaining, that is when it is able to supply itself with sufficient resources, whether energy or materials. To be truly sustainable requires both. Such a requirement is likely to be too strict, since neither solar energy nor material resources are distributed equally in all parts of our world. An immediate demand emerging from the wish of achieving sustainability is that, before developing the proper strategies we are able to prioritize between initiatives and make the best, most sensible choice. This requires that we should be able to evaluate between the many different forms of energy and (re)sources of materials on which our society is based. One physical property allows this to be done, namely the concept of work energy (exergy). Thus, this concept was chosen to be used as a common denominator in the description of all pools and processes in our society. So, when talking of energy in the following, we will always address problems in terms of exergy, or as we rather should formulate it: work energy, since it describes the ability of energy to do work, both physical and chemical. In the project, we deal with the identification of areas of high consumption or unnecessary losses of work capacity that might be reduced with relative ease or used in a more sensible manner.

It must be noted that the normal way of doing energy accounting does not permit such an identification as it does not distinguish between the different quality forms of energy. It serves merely as a trivial accounting system following the first law of thermodynamics, which in the end says very little about how well a given system actually performs.

The new method has been developed so that it may be used not only for the identification of target areas but also as a monitoring device so that a transitional process may be followed and eventually steered in the direction of increasing sustainability of a society. One basic idea was that it should be possible to convert the data collected in a particular format from a given web page and get an immediate calculation of the indicators both for various sectors and for the society as a whole.

During this project, a great deal of attention has been paid to ensuring that the approaches taken during the implementation of the methodology would also ensure a high degree of universality, in the sense that the method developed would take into account the "systematics" already established within the European Union as well as by various Danish authorities. Thus, for instance, the method developed should reflect the ordering of landscape as described in the Corine Land Cover classification system (www.eea. europa.eu/publications/COR0-landcover) and be compatible with this, as well as being easy to integrate with the Geographical Information System–based reporting systems used by farmers for the location and identification of crops.

Following identification of the important work energy locations in a system, it is important to evaluate how the system is doing in terms of other parameters normally used as indicators of environmental performance. For a society, this could for instance be done by establishing carbon budgets for relevant processes and determining the CO_2 emission per capita. The framework thus established should also make it possible to foresee the consequences of possible measures that might be taken in the future, that is make it possible to analyze scenarios. Therefore, along with the development of the work energy analysis, the data available from this can at the same time be used to feed an additional module that estimates the consequences to the carbon budget of the societal system during the transition or to make studies of various scenarios (see Jørgensen and Nielsen, 2015).

The project has been followed by a Project Reference Group consisting of Dr Simone Bastianoni, University of Siena; Prof Per Christensen, Aalborg University; Andy van den Dobbelsteen, Technical University of Delft; Prof Michael Hauschild, Technical University of Denmark; Dr Felix Müller, Christian Albrechts Universität zur Kiel; chief consultant Sigurd Lauge Pedersen, Energistyrelsen; and Dr Sven Stremke, Wageningen University and Research Centre.

Preface

The earlier mentioned idea of an optimized function of nature started out with a desire to improve existing models in order for them to be able to reflect natural properties, that is to allow them to include processes such as adaptation, buffer capacity and the ability to perform transitions between species, changes in ecosystem network and so on. Such phenomena, although normally observed in nature, had not previously been included in the "state-of-the-art" bio-geochemical models of the 1980s, even though it was widely accepted that such properties were general and commonly possessed by all natural systems, certainly by organisms, and possibly even by ecosystems. The hypothesis was that the result of incorporating such features into the models would result in a better fit between model simulations and actual observations (see Section 1.7).

Behind such an effort some other assumptions were hidden, namely that ecosystems during their development should evolve in a manner which would exhibit a directional behaviour, where one could forecast certain trends in the number of components and interrelations, energetic efficiencies and so on (Jørgensen and Mejer, 1979, 1981; Mejer and Jørgensen, 1979). Much later, the idea emerged that both ecosystems and societies were complex systems that might be evaluated in the same manner. If we could justify such a view, it would mean that we could probably benefit from comparative studies between ecosystems and societal systems. In turn, we would improve our understanding of both types of systems by studying the similarities and divergences in the performance patterns of the two. This is partly what this project demonstrates, although many of the principles applied to a society here have been developed on ecosystems first.

The previously described systemic and inherently physico-mechanical approach has not been implemented on ecosystems without obstacles. There are many reasons for this. First of all, there will be some technical questions such as "How to interact with the model?" and "Is it possible to interact continuously, which is problematic because it will be time-consuming, and if we settle for discrete interactions, then how often will we need to interact with the model to mimic the processes of adaptation and selection?" The answer to the latter question will probably need to be a compromise, since the key components of ecological models—the state variables—often operate on very different time scales. How strong do we want the interaction to be? And—for instance—how quickly and how much can organisms adapt? Answering these questions involves quantitative and qualitative aspects of the system modelled, as well as raising programming issues, a case of modelling Complex Adaptive Systems (Holland, 1992; Abel, 1998).

The second problem seems to raise an even more complex issue. In order to carry out such an intervention in the model, we need to take a stance on what exactly governs the evolution of an ecosystem. In what direction will organisms adapt their parameters in order to respond in the most "sensible" way to changes in their environment? What rearrangement or replacement of organisms will take place? Which direction will the system take as a whole? Will it be the same direction throughout all its evolution? Will one descriptor be able to serve as a common, unified indicator of the ecosystem state and its performance?

The introduction of such a view of directionality (telos) in nature has been viewed as heretical ever since Aristotle, and it has only recently and ever so slowly been adopted again, for instance in disciplines of a new theoretical ecology and biological semiotics (Jørgensen, 1992). Nevertheless, technically speaking, when testing the previously described hypothesis a guiding algorithm—in mathematical language: a goal function used to govern the evolution of the system—needs to be formulated and incorporated into the model. The mathematical term *goal function* says all there is to say. We easily get back to the question about the acceptance of a telos in nature. But in this case the telos is formulated in the form of a strictly materialistic-based metaphysics, and the goal function is only used to steer the evolution of the model and not necessarily to lead to an end solution where the function has also been maximized. For a discussion of this, see for instance Nielsen and Jørgensen (2013).

Thermodynamic functions have been tested for some years now, and all in all there are strong indications that such a type of function does have a meaning in an evolutionary context. This is not to say that we have a full picture and are able to say exactly how this pattern comes about. Put briefly, what is most striking is that all the mathematical expressions used as indicators in theoretical biology as well as in ecology seem to come from a "family"—a set of similar expressions—a phenomenon which in mathematics is referred to as isomorphism. When viewed from the outside, there is only little difference between the widely used (bio)diversity measures and more technical expressions such as the Boltzmann–Gibbs equation and the Kullback–Leibler measure of divergence (Kullback and Leibler, 1951)—to mention but a few examples (Mejer and Jørgensen, 1979).

Nature—or rather its compositional elements together with their functionality and the efficiency of their processes—has by now been tested over a period of about 4 billion years. So we may easily assume that whatever we observe or encounter in the organisms and ecosystems around us has been thoroughly tested over quite some time. Therefore, the existing structures represent mechanisms and functions that are robust, that are able to meet a vast variety of changes over time. The only question left seems to be if it is sensible to apply this conclusion to the ecosystem level as well. Last but not least, we must raise the question as to whether such principles are also valid for societal systems. Will they tell us anything about how well we perform with respect to being sustainable?

Along the path of working with such ideas, some initial and probably essential observations have been made. For many of the indicators used in studies of ecosystems, it is found that when applying the same measure to the study of societal systems, and on the same premises, the behaviour seems to be quite contradictory to what is generally observed to occur in an ecosystem. The things observed to happen in society simply do not reflect how nature would have handled it; that is they do not represent viable solutions in nature (e.g. Nielsen, 2007; Nielsen and Müller, 2009). In other words, the ways we govern our society lead to a conflict with the observed behaviour and evolution of natural systems or ecosystems. Much of this can be explained by the interaction between our excessive use of resources and the economics of this "cradle to grave" use of resources, as well as by the almost total lack of efficient recirculatory processes in the systems—as also illustrated in this presentation. Studies of the interactions between these mechanisms and their importance to the improvement of environmental conditions are the focal point of recent scientific disciplines such as ecological economics.

From these two queries arise: What will actually be the result if we adopt the same viewpoint in order to study the behavioural evolution of societal systems and to evaluate the health/efficiency of processes in them? and will such studies bring us closer to understanding and defining sustainability?

The present book represents an attempt to combine and use ecological as well as engineering knowledge in order to establish a framework and a methodology within which we can interpret the degree of sustainability of societal systems. It is believed that establishing such a framework will assist in defining the necessary and sensible target areas towards which we can steer future developments, thus finding the optimal direction leading us faster and better towards a truly sustainable state (see Chapter 3).

The approach encompasses one major focal idea, that is that all activities taking place in our society can in some way be connected to a spatial component in addition to a temporal variation. That is, that all societal structures, their organization and pertaining processes can be connected to a given point in space and time. In the simple case considered here, we mainly work with two dimensions only. This means that measures of activities as well as their impact can be estimated from and related to information that can be derived from any sort of Geographical Information System and therefore are not dependent on specific software packages. Thus, data for municipal parameters, agriculture, nature, and so on have to a large extent been derived from the MapInfo files available at the administrative offices of the municipality; this is the software used to store or extract geographical information about the island.

To evaluate the overall performance of a society in terms of sustainability, it is important to be able to identify specific sectors where the situation may be most critical in determining the overall state. Therefore, as seen in the report, our society has been divided into a number of fundamental units of activity thought to be present and relevant in almost all societies. Thus, the analysis has here been concerned with the public services, private life and existence, agricultural production, services and industry—and finally the importance of a properly functioning nature *sensu stricto* able to supply the ecosystem services necessary to the survival of the society.

Why choose an island for this first attempt at developing and applying a new methodology? A simple, first answer is obvious since an island offers a system—containing a society as well as nature—included in a fixed and controllable boundary over which all important flows of energy and matter must pass. This represents a condition that was assumed to be necessary for a first try. At the same time, it is obvious that such an "isolated" system may in fact be too simple to illustrate all the complex functions of a full society. Meanwhile, to as high a degree as possible, the method has been formulated so that at least a framework has been laid out that makes it feasible to upscale to larger municipalities, regions and even countries.

Acknowledgements

The author is grateful to the VELUX Foundation for its support for this project, which was crucial to this type of project where new methodologies need to be developed and the major problems in this development have to be identified. I am also indebted to the people on the island of Samsø for sharing with me many experiences and information about life on the island. In particular, I want to thank the Energy Academy for housing and administering this project: Søren Hermansen and the staff under his leadership, a staff that was always helpful and willing to share their network with me, making it possible to cover almost any issue that arose. Also, I have been lucky to have my colleague through many years Sven Erik Jørgensen with me along the challenging path of development of his and Henning Mejer's ideas of exergy maximisation or optimisation in ecosystems, a process in which I have participated since the mid-1980s. The idea of the existence of such a feature in ecological systems, being the underlying cause of the behaviour and development of nature, led almost inevitably to the obvious question: What happens if we view our societal systems in the same way as we view natural systems? This book contains the first partial answer to this question. Finally, I wish to thank my long-time friend and fellow in singing, MA and Cand. scient. Mike Robson for investing his native language skills in improving this text.

Søren Nors Nielsen

Author

Søren Nors Nielsen earned a PhD in Structurally Dynamic Modelling in 1992 at the University of Copenhagen and in 2009 he achieved the senior doctorate degree of Doutor agregado in ecology at the University of Coimbra. He is a world expert on the application of thermodynamic methods for the analysis and assessment of biological systems functioning, mainly in aquatic environments, and of ecological modelling in environmental management. He authored or co-authored a large number of papers in peer-reviewed journals, book chapters and books.

Reading Instructions

The chapters take the reader through a sequence which covers some major problems within the field of sustainability, such as the "currency problem" (comparing energy, matter and money) together with the evaluation problem (how to determine ecological value within the framework of a traditional growth-based economy). Some basic considerations have been set out already, and some more details are explained in Chapter 1.

The concept of work energy or exergy is introduced in Chapter 2 together with a brief explanation of the advantages of using this concept, primarily (1) that it overcomes the "currency problem" between energy and matter directly by reducing them to the same unit, namely that of work energy (Joules, Watt-hours or similar units), and (2) that work energy analysis is a much more powerful tool than traditional energy analysis for determining focal points that should be subjected to more intensive studies and indicating where action can be taken and measures applied with the largest benefits in terms of increasing sustainability.

The methodology for a possible implementation of such an analysis in a relatively small society possessing a fairly well-defined boundary is presented in Chapter 3. Here also the basic background to the thermodynamic analysis of a society is established, and suggestions are made for a monitoring system and for the calculation of some sustainability indicators. The chapter introduces a sectorial division of the society into parts that are found to be essential for its function, together with a presentation of the first basic data: a geographical estimate of the area occupied by the various sectors.

For those interested in results only—either because they already accept the methodology or because they are not interested in struggling too much with the theoretical background—it is possible to start reading after this chapter, although with the warning that the terminology developed and advocated in the first chapters will continue to be used throughout the rest of the text.

As many Danish municipalities are currently monitoring their energy budgets, the method is initially applied to such a budget for the study island of Samsø. In principle, this is the traditional way of analyzing energy flows in a society, and it follows the first law of thermodynamics. In its present form, the conversion scheme is adapted to the format of the data in municipalities where the energy budget is made by the company PlanEnergi. In order to achieve the new level of a second-order analysis, the conversion scheme then needs to be expanded by introducing a quality perspective to the different types of energies in it. The overall energy budget (first law) and a first simple conversion to work energy (second law) are found in Chapter 4.

It should be noted that in the established some redundancy of terms seemingly has emerged. This redundancy is intentional, and the implemented schema has been kept to ensure stringency. Furthermore, some of the terms that are at present void, that is have no presently known value, may in the future suddenly become important. Thus, a general feature is that, presently, the terms involving storage of energy are not important in this analysis as we have only inferior solutions today (Chapters 3–9). Filling out these void spaces in the analysis may become an important part once proper technological solutions have been established.

The studies, data collection and results of various sectors are described in the subsequent chapters: Chapter 5 describes the *public sector*, and Chapter 6, the private *household sector*. Chapter 7 addresses the *agricultural* activities, mainly split into crop farming and livestock. In principle, this sector would also potentially include fisheries, an activity which could be important to other societies. Fishing was previously carried out on the island but was found to be no longer significant in connection with the present study. The industrial sector on the island is limited and mainly consists of production activities related to agriculture. In addition to these processing activities, the sector also comprises activities related to trade and other commercial activities. These are all presented in Chapter 8.

Nature is described in Chapter 9, not as a societal sector per se, although it is believed to represent an important activity, servicing our society through what are nowadays recognized and known as "ecosystem services". The accounting of this "sector" follows very much the landscape grouping described in the Corine Land Cover Code system (see the earlier discussion). The approach presents estimates of both a more traditional work energy (derived from thermo-chemistry) together with a calculation of the new concept of eco-exergy, which also aims to include the information content of the respective components in the ecosystem as derived in Jørgensen et al. (1995) and Bendoricchio and Jørgensen (1997). In the end, the methodology presented here settles for a comparison which does not include the information part of nature; this is considered to be the fairest solution as no methodology for the inclusion of the analogous information contents in our society has yet been identified. It is striking that we know so little about quantification of the importance of information to the structuring and functioning of our society.

As a side product which will probably become important in the future, an estimate of the work energy content of wastes has also been made. The topic of wastes is considered to be a major focus area as the islanders have decided to make an attempt to establish a circular economy on the island—part of a plan known as Samsø 3.0. Producing such an estimate was not originally foreseen, and it remains at a rather immature stage, but it reveals the potential inputs to the idea of running at least part of the island's transport system on some sort of biomass-derived fuel (Chapter 10).

The final chapters attempt to come up with some conclusions for the future sustainability of the island—Chapter 11—including a discussion of potential improvements and necessary steps to be taken if the system is to be implemented elsewhere. The chapter also addresses concerns relevant to the upscaling of the system and its methodology—challenges that are being tested at the moment in the ongoing project in Jämtland. Finally, the chapter includes a hypothetical scenario of a situation in 2020 on the road to a society independent of fossil fuels.

A series of conclusions have been drawn in Chapter 12, derived from the method development, its implementation and the subsequent results.

Summary

For some decades now, the concept of work energy has been used as an entry point to studies of flows of energy and matter in nature and ecosystems. By such studies, it has been demonstrated how the thermodynamic relationships derived from the flows and stocks of energy have great importance to ecosystem function. It has become an obvious idea to make an attempt to use the same viewpoints in order to study and improve the functionality of our societal systems.

From time to time, it has been ascertained that nature and society—even though sharing a series of essential features—seem to develop in quite different ways. Both types of systems are complex; they are hierarchical constructs; their constituting components exchange energy and matter; they are composed of networks—and even so they perform so differently. What are the reasons for these differences? And would it be possible to use a deeper insight and understanding of this question to achieve insights in terms of sustainability issues?

Nature consists of self-organizing and seemingly also self-optimizing systems—but what about our societies? In many ways, it seems clear that our society exhibits some developmental trends which are in direct contradiction with—and sub-optimal in comparison to—what would be observed in a natural system. Is it possible immediately to conclude from this that societies are poorly constructed—or is it merely our insights and understanding of society as a complex system which is inadequate?

If the previously mentioned studies can contribute in a sensible manner to an understanding of the functional principles, development and evolution of nature—would it not be interesting to employ the same principles to analyze a society to find out whether or how such an implementation could contribute to an improved understanding of the function of our societies within the context of sustainability issues? It is such considerations which lie behind this project.

As a first attempt, it is convenient and desirable to work with a clearly demarcated system. In this respect, an island is useful as a model for a society. However, the island needs to be of a sufficient size so as to comprise most of the functions exhibited by a society. The demarcation of the island to a certain extent makes it easier to account for activities that may reach out over its boundaries whether imports or exports.

The Danish island of Samsø was chosen as a model society. Since 1997 the island has been re-organizing its energy supplies as part of a "Sustainable Energy Island" project. In 2005, the island managed to achieve self-sufficiency in energy for electricity and partial self-sufficiency in energy for heating. This appears from the energy accounting reports elaborated every second year by the company PlanEnergi. For the development of methodology and analysis, the year 2011 was chosen, since at the time when the core studies were carried out this represented the latest year for which an energy accounting report had been produced.

First, a normal energy budget accounts only for the flows which we normally perceive as energy and not for flows of materials. However, part of the accounting comprises energy which is bound in materials such as fossil fuels or biofuels. To complete the picture of work energy distribution in the society, it is necessary also to describe and quantify the materials and material flows on the island.

In this manner, work energy (exergy) becomes interesting as a concept serving to unify the two types of flows that are necessary to drive our society, that is, on one hand, the work energy related to the production and consumption of energy as such and, on the other hand, the work energy incorporated in materials, chemicals and products. The latter type of work energy forms part of most of the things surrounding us in our everyday life, from our houses to the production apparatus supplying us with consumable goods of all kinds.

Second, normal energy budgets also ignore the point that some of the energies are no longer at our disposal. The energy passing through our society, and being transformed by the various processes undertaken within it, tends to lose its most important quality, namely its capability to do work. This principle should be our main focus of interest, and the preservation of this capacity should be our target in optimizing our society to bring it into a state of sustainability. As work energy is expressing this property, it has been chosen as the core concept for the development of a stringent method for evaluating the level of sustainability in a given society.

The methodology has been developed to be as general as possible; that is in principle, it may be applied to any society, although it is foreseen that some work must still be invested in adapting it, for instance in connection with an upscaling that brings other sectors or production activities into play. However, with the methodological foundation laid out here, this work is considered to be limited.

The method is built up in a hierarchical manner so that the society has been divided into a number of sectors and, in a few instances, also sub-sectors. These may all, in turn, be subjected to further analysis. In the case in question, the society has been divided into six sectors: an energy supply sector; a public sector; a private household sector; an agricultural sector; an industry, trade and commerce sector; and nature. Wastes, and solid waste, in particular, were not originally planned to enter directly into this analysis. However, it turned out that quite detailed data were available for this area. Therefore, an initial tentative analysis was performed to evaluate the potential work energy involved and the potential role of this area in the future. This is particularly interesting when considering the plan to be free of fossil fuels in 2030.

In establishing the method, a specific starting point was adopted, based on the fact that most of the activities undertaken by our society can in one way or another be related to a physical geographical area. Cities, houses and roads, agricultural activities, airfields and nature are all entities taking up physical space and to which we can ascribe a certain area. To ensure the general aspect in the description the European Corine Land Cover Code (CLC) system has been used for a start (EEA, 2007). The CLC codes are to a wide extent reflected in the format of the data held by most Danish municipalities, usually in the form of files in a Geographical Information System (GIS). These data, together with information on buildings and building sizes (Danish Bygnings- og Boligregistret (BBR) and Ejendomsstamregister (ESR) registries), form the basis for an estimation of infrastructure and partitioning of activities within the landscape.

Quantitatively, the public sector is relatively small, and in work energy, accounting it appears to be quite inefficient. The reason for this is that production proper, for example the manufacturing of goods, is not the purpose of this sector. Its role in society is to provide us with the non-material goods and services necessary to run our society.

The private sector likewise appears rather inefficient and is represented by a relatively large amount of infrastructure, namely the houses we live in. These are also places in society where a lot of the individual consumption of goods occurs, and goods are also stored here on varying time scales. Unfortunately, it has not been possible to get reliable estimates for this part of the sector. As with the public sector, the purpose here is not production. Indirectly, the size of the sector ensures that a considerable quantity of renewal and repair activities will create corresponding activity in other sectors of the society—thereby contributing to other forms of sustainability.

The agricultural sector plays a relatively large role in determining the appearance of Danish landscapes and on Samsø activities related to this sector take up three quarters (75%) of the island's area. The individual farmer reports to governmental systems what areas ("field blocks" in Danish) in the landscape are used, their respective areas and what crops are grown on them. The crops are identified by yet another code system. The data for this have been obtained from the home pages of The Danish AgriFish Agency and Ministry of Food, Agriculture and Fisheries of Denmark.

The industrial sector will vary a lot between different types of society. Due to the rather limited extent of this sector on Samsø, this is likely to be the area where the greatest adaptation of methodology and data are needed if the method is to be used elsewhere or on a larger scale. In the case of such an upscaling, it is suggested to take a starting point for the methodology in the inventories of factories which already produce some sort of environmental accounting, have implemented systems for environmental assessment (EMAS) or are already certified, for example according to the standards of the ISO14000-series.

In the specific case of Samsø—although the sector appears rather limited—the infrastructure accounts for a relatively large area. Similarly, the consumption of fossil fuels is comparable to the amounts used in the private household sector and agriculture.

The amount of work energy contained in nature—expressed in terms of the chemical energy of its biomass—is similar to that of the cultural sectors. If the concept of work energy is extended to also comprise the information held within the organisms of the ecosystem, the importance of this sector becomes overwhelming. This information is not passive but plays a rather essential role in maintaining the functionality of the ecosystems. This only serves to emphasize the real value of nature, not only in itself but also when we enjoy all its varieties of "ecosystem services". The information content is fundamental to this function.

The partitioning into sectors required the unravelling of more data for calculation of the importance of particular activities and their respective and corresponding use and consumption of work energy. Detailed knowledge about societal activities is not always easy to obtain and is sometimes associated with high levels of uncertainty. In most cases here, the contributions of items that could not be quantified in the overall budget in terms of work energy are assumed to be so small that a more accurate determination will not affect conclusions significantly.

In general, a large amount of work energy is bound up in the infrastructure, primarily buildings and roads. Even when using cautious estimates of the work energy needed for a continuous renewal of this infrastructure one reaches values that call for increased attention to be paid to this aspect in the future.

The transition to energy self-sufficiency seems to have been accomplished now, at least with regard to the energy supply sector. However, considerable amounts of fossil fuels are still being imported to the island. This input of work energy must be replaced in one way or other if the island is going to be free of fossil fuels. Most of the imported fuel is used for heating and transport. Both types of consumption could to a wide extent be replaced by utilizing the surplus electricity production—even with present technologies.

Considerable amounts of work energy are exported from the island, mainly in agricultural products, manufactured or fresh, together with their associated packaging materials. At the same time, the majority of the island's solid waste also leaves the island. The materials are considered to contain an amount of work energy which could be advantageous to use on the island. The excess production of electricity is likewise exported but could in the future be diverted to other purposes on the island—contributing further to its sustainability.

Overall, it can be concluded that the island of Samsø through its investments in sustainable energy has taken a great step forward towards becoming a sustainable society. The island still has a number of unexploited resources which potentially could replace a major part of the fossil fuels still being used. It is estimated that these resources, together with the excess production of electricity, represent a platform for local energy production that may reach a level that brings the island close to full sustainability.

1

Introduction to Sustainability and Work Energy Analysis

1.1 Introduction to Sustainability Analysis

In this chapter, we investigate the currently accepted definition of sustainability and identify the need for a more precise definition, since only a unifying approach with clearly specified requirements can properly assist us in constructing a sustainable development process. In short, the moral obligation to care for future generations as laid down in the Brundtland report (WCED, 1987) and Rio Convention (UNESCO/UNCED, 1992) is clearly insufficient, for instance when formulating appropriate political actions to prevent potential damage from whatever we will face of changes in our environment arising from forecasted climate changes. An idealistic wish does not provide guidance for specific actions. The recent introduction of sustainable development goals (SDGs) represents a step forward (UNDP, w/o date), but again, a long list of possible initiatives will not necessarily cause the proper measures to be taken.

Many attempts have been made to sharpen the definition of sustainability, in particular the problem of how to integrate the various proposals in a more systematic way. This includes considerations on systemic properties and approaches to the analysis of sustainability in regions and society as well as in nature. Interestingly, presentations of the concept of sustainability often give priority to economic and macroscopic social issues, whereas—as this study shows—the world looks different when viewed through the eyes of laymen and ordinary people. Suddenly, the local, microscopic and environmental issues come into play.

The dichotomy between our perception, recognition and eventual conceptualization of problems affects possibilities and restraints in the future and thereby represents a key issue in our attempts to achieve sustainable development. Such issues need to be resolved. The chances of achieving progress and moving society towards increasing sustainability seem to be higher the closer scientists manage to work with local people who experience problems directly in their own lives.

One special problem remains to be solved in all theoretical aspects of achieving sustainability. In fact, the issue is very simple when dealing with types of societal analyzes which are able to tell us simple things such as what we are doing and what is the relative importance of the things we are doing. Such simple knowledge is necessary in order to tell how close or far away from sustainability we are and what we can do now or will have to do to in the future. Among the many approaches proposed in the literature during recent decades, work energy (aka exergy) provides us with a strong candidate to use as the entry point for such an analysis. All in all, it is considered to provide a satisfactory unifying platform that enables us to make a cross-boundary analysis comprising both energetic-material and societal and environmental issues.

Ever since the publication of the report of the Club of Rome, widely known as "Limits to Growth" (Meadows et al., 1972), there has been public concern about our excessive use of natural resources and their possible depletion. This book appeared only shortly after the author Rachel Carson in another publication, *The Silent Spring* (Carson, 1962), had pointed out the negative consequences of our industrialized activities and production methods arising from their utilization and release of various types of chemicals and wastes which were having a high impact on the state of our environment—which, in turn, would have an impact on ourselves and our societies. We have to learn much faster from our mistakes (EEA, 2013) and pay much more attention to the potential dangers inherent in our activities by applying responsibility (Jonas, 1984) through the precautionary principle (Kriebel et al., 2001).

It has become clear that depletion of the primary energy drivers of our society, namely fossil fuels in the form of oil or coal, is coming closer and closer to its limit; it is only a matter of time before the situation begins to impact the present level of global activity. Nevertheless, we face great difficulties in developing a concerted plan of action that would allow the necessary measures to be taken to counteract this development.

Meanwhile, the situation with fossil fuels finds its parallel amongst almost all the material resources that we regard as necessary to the way we currently live our lives. As pointed out by Hubbert as early as 1962 (Hubbert, 1962), most resources will sooner or later be exhausted. This is indeed the situation not only for many of the metals we use but also for simpler materials like sand for cement, and plaster for the construction industry.

Concerted actions may in some cases be much easier to take on a more local scale, as is the case here with the island of Samsø, where the obvious motto has been to think and act locally.

1.2 The Sustainability Concept

The concept of sustainability appeared in our language in connection with a report published by the World Commission on Environment and Development with the title *Our Common Future* (WCED, 1987) also known as the Brundtland report. The publication ably represented the outcome of a much longer process initiated in Stockholm in 1972 where the UN held a conference concerned with the human environment. This was at a time when the issues raised in the previously mentioned books by Carson and the Club of Rome were already well known. Some 5 years after the appearance of the Brundtland report, an international conference was held in Rio de Janeiro, resulting in the so-called Rio Convention or Rio Declaration (UNESCO/UNCED, 1992). Ever since it first appeared, the definition of sustainability has been criticized for being unclear, and many thoughts and suggestions have been dedicated to the questions, What does it actually mean to be a sustainable society?—and, if we can possibly agree on a proper definition, How do we actually achieve it?

The most commonly quoted citation and also a very central formulation of the issue of sustainability comes from the Brundtland Report (WCED, 1987), which tells us that

> [s]ustainable development is development that meets the needs of the present without compromising the ability of future generations to meet their own needs.

The difficulties caused by such an imprecise statement have attracted much criticism. What exactly will our needs be in the future? In fact, do we even know what we need at present? And is it necessary to meet our needs in the same way in the future as we do today? Would it be sustainable to use up some resources to create a new technological platform for future existence? After all, we have put some severe constraints on future generations by using up most of the easily accessible fossil-fuel deposits on Earth in just two centuries. Wouldn't it be better to meet the "needs of future generations" by being extremely foresighted? This means that we should start to develop new technologies to replace the existing ones in order to substitute non-renewable resources with renewable ones. Considering the severity of the situation, many consider that this cannot happen fast enough.

There is no doubt that without a much clearer and more stringent definition, sustainability stands out as a rather vague concept. The different angles from which the topic can be addressed—economic, societal and environmental—leave many people with the impression that we can settle for only one of these aspects. In fact, we need all of them at the same time. That additional flavours have been added to the concept so that a society can be designated as possessing either weak or strong sustainability (Daly and Cobb, 1989; Barbier, 2019; Barbier and Burges, 2017)—and even different graduations between—does not help the situation.

Although the Brundtland report comes out with some quite clear descriptions both of the problems and of the actions needing to be taken in consequence, it is not clear how and where these actions should be taken. Who is responsible for their implementation?

More recently, several new concepts and ideas have emerged, and these are now proposed as a possible way to integrate and evaluate the cycles of energy and matter flow that occur within our societies as well as in nature. Work energy represents such a concept. It finds its strength in its applicability to a wide range of environmental issues, thus serving as a common platform of analysis. This is considered necessary in order to get closer to a more precise definition of what sustainability actually means to our present and future societies.

1.3 Sustainability—A Materialistic Definition

A major and fundamental issue concerning the resources that we use to drive our society is that they mainly come in two forms, either energy or matter, and both forms basically stem from resources that may be considered either renewable or non-renewable in character (see Figure 1.1).

As mentioned earlier, the double-dual character of the work energy flows is adopted as the entry point of this analysis. Thus, a given flow will be assigned a position on a sustainable–non-sustainable axis and, at the same time, be designated as belonging to either an energy or a matter flow (see the end of Chapter 3) (see Figure 1.2).

In this context, we distinguish between sustainable and non-sustainable flows or resources by correlating sustainability with renewability. Simply formulated, a resource is considered to be sustainable when it is renewed (formed or regenerated by reuse or recycling) on the same time scale at which it is consumed/used by the system (see also Section 1.3.2).

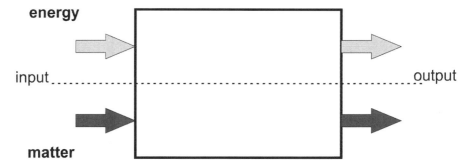

FIGURE 1.1
In many representations of our society the resources used to drive our activities have been shown as either belonging to flows of energy or matter. Both are used to produce products, goods or services we believe are necessary for the realization of our everyday life (e.g. Cleveland and Ruth, 1997).

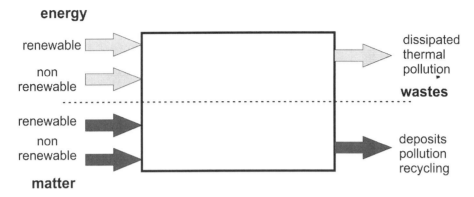

FIGURE 1.2
Inspired by H. T. Odum, a further partitioning of the two flows is undertaken. To reach a sustainable state of our society, it is of utmost importance where the energy or materials are coming from, whether they stem from non-renewable or renewable resources. True environmental sustainability is reached only when all energy is based on renewable energy forms and all materials are perfectly recycled and extraction of virgin materials is not needed.

1.3.1 Energy Perspectives

We need energy to run our society, and this energy is derived from sources that are very different in character: these include fossil fuel, biomass, solar radiation, wind power and hydropower.

Also, the extraction of energy from these different sources requires very different technological levels in a given society. A wood-burning stove for domestic heating differs significantly in its technological demands on society from domestic heating solutions using district heating or co-generation plants—which demand an advanced infrastructure—or from solar panels, electric heaters or heat pumps.

A huge problem is presented by the fact that a major part of all energy used today is directly derived from finite resources, that is resources that if not recycled will run out someday. Not only resources such as coal, oil and natural gas but also tar sands and shale oil belong in this category—although on different time scales—and will eventually turn out to be finite. The same is the case with energy stemming from nuclear fission plants, not considering the various other problems, such as the residual radioactivity in the wastes, arising from this type of energy production. For interpretations emphasizing the role of thermodynamics, refer to Rifkin (1989), Cleveland and Ruth (1997), Daly (1992), Hammond (2004).

According to some authors, we may be able to reach a situation where we are able to exploit a much larger proportion of the solar radiation reaching the Earth (e.g. Ayres, 1998, 1999). This optimistic view is built on the fact that the energy input from solar radiation in principle is unlimited. We just need to become much better at using it, and then all problems of energy supply will have been solved. But in the meantime, the improvement of such exploitation is not only dependent on existing technologies but also closely linked to material resources that are finite.

The discussions on such issues originate in the works of Georgescu-Roegen (1971) (Daly, 1995; Gowdy and Mesner, 1998) and have continued ever since. As Georgescu-Roegen pointed out, no energy conversion will be perfect, and all conversions of materials too will lead to dissipation (see the later discussion). His conclusion is therefore that once energy has been invested in upgrading matter (as illustrated, for instance, by decreased statistical entropy; Rechberger and Graedel, 2002), we should do everything within our power to keep it at an upgraded level and save energy by not allowing it to disintegrate and spread again (see Section 1.4). This point accentuated the value of reuse and recycling of materials at a time when this had not yet become an issue and demonstrates how foresighted the views of Georgescu-Roegen actually were.

1.3.2 Material Perspectives

All material resources seem to share the property of being finite; by comparison, see Hubbert's peak (Hubbert, 1962) described earlier. The amount of matter available to us is limited by the size of our planet and by the size of the part of the Earth's crust that we are technically able to exploit. Finite in this context means that these resources will eventually come to an end and that the extraction of virgin material cannot continue forever. For convenience here, we will not enter into the discussion of matter being available as either resource or reserve and whether these have actually been proved to exist or not. The issue of matter existing as either a resource (total) or as a reserve (economically viable to extract) involves economic considerations that are beyond the scope of this treatise.

If finiteness is considered, a resource can only be considered as sustainable if it is not consumed at a rate faster than the rate with which it is (re)produced. By production, we here mean that the resource is produced so it may be utilized as virgin material. By reproduction we refer to the fact that reuse and recycling make it possible to use the same materials several times, thus reducing the pressure on and demand for virgin materials.

1.4 Integrating with Physics

One of the major obstacles to giving a clear(er) definition of the concept of sustainability—as almost seen from the definition given earlier—is that such a definition must necessarily involve a set of paradigms that do not normally share a sufficient number of world views to make this an easy task.

Getting closer to a proper and rigorous definition must necessarily involve disciplines such as the following:

- Physics
- Chemistry
- Geology/geography
- Biology
- Ecology
- Sociology
- Economics

Of these, the first five are viewed as fundamental elements of environmental sustainability, without which the sociological and economic versions of sustainability will not exist. Hence, the latter are seen as inferior and not included.

Many intertwining subdisciplines, that is mixes of knowledge from the preceding discussion, such as atmospherics, meteorology, hydrodynamics, physical chemistry, thermodynamics, soil sciences, knowledge of climatic belts, organisms and populations, must be brought in before we can determine an interaction with our society, its infrastructure(s) and eventually economics.

The definition must, on one hand, be wide and involve multi-, inter- and transdisciplinary approaches. The problems will not be solved by one scientific discipline alone but, rather, by the application of several approaches that span several of the mentioned disciplines. Meanwhile, on the other hand—as stated earlier—the fundamental relationship to and dependency on resources, both energy and materials, must be recognized as the key issue in achieving sustainability, thus giving rise to a demand for what is known as "the supply side of sustainability" (Allen et al., 2003).

In many ways, as we shall argue through these first two to three chapters, the science of thermodynamics—even if applied on an intuitive basis— provides a good candidate for a common interpretational platform which may serve the purpose of facilitating the interaction and translation (mediation) between the disciplines.

1.5 Thermodynamics and Society

The approach of imposing a thermodynamic framework onto the interpretation of societal function in order to improve our understanding of the interaction between environmental management, nature's working principles, society and economics has a long history and will only briefly be described here.

The introduction of a thermodynamic view of society and economics in particular has previously been proposed by the economist Georgescu-Roegen (1906–1994) in a seminal book titled *The Entropy Law and the Economic Process* (Georgescu-Roegen, 1971). Although he was born in Romania, much of his research was carried out in other countries such as France, England and the United States.

Although his ideas set out the framework of the new subdiscipline of ecological economics, his approach and viewpoints never became sufficiently popular to take over the political agenda in the debate between environmental policy and economy. This fact is probably due to conflicts between established paradigms; not only was he conflicting with several core concepts of economy—which was serious enough—but many physicists reading his papers and books also found the physics he presented not deep enough for them to take the approach seriously.

Above all, thermodynamics is commonly viewed as a very complex and difficult science. Attempts to popularize the understanding of for instance the impact of industrialized countries as entropy-forming structures, as it happened in the works of the science journalist Jeremy Rifkin in his book *Entropy, a New World View* (Rifkin, 1989) seem not to have had sufficient impact for it to be noticed by the vast majority of the public, to say nothing of managers or politicians.

Much focus has been put on the destructive or dissipative side of the evolution of systems, spanning over a wide range from mere physical processes to self-organized non-living structures—via simple life-sustaining systems, cells to organisms and ecosystems and eventually to our societies. This is simply because all activities have a price, namely the loss of the capability of energy to do work, expressed through the formation of entropy. Meanwhile, we must not ignore that we pay this price precisely in order to create, improve or maintain the structural components of our society. To confuse things even more, the classical debate around possible connections among entropy, information and order/disorder has also been involved here. Nevertheless, this debate is ignored here, and the discussions that follow are based on a strict physico-chemical interpretation of energy and entropy.

To add even more to the confusion, two classical but very different approaches to the importance of thermodynamics in the evolution of systems exist, namely that of the "minimum dissipation principle" of Prigogine and co-workers (popularized in Prigogine and Stengers, 1984) and the "maximum entropy production" (MEP) principle of Jaynes (1957a, 1957b) and Swenson (1989). Although some of the issues between these two seemingly contradictory approaches may quite easily be explained and solved by addressing differences between entropy as an extensive vs. intensive variable, or even differences in semantics, we will not be concerned with these problems here, only indicating that the controversy exists.

So in spite of these efforts, there seems to be only little interaction between the entropic views of society and economics, and McMahon and Mrozek (1997) state the issue to be that this view is "irrelevant in the neoclassical view, relevant in the ecological". Meanwhile, the continuation of the paper points to thermodynamics as an important constraint to the economy ("economic growth" Sic!), and this is the point to be followed here, as already mentioned.

Nevertheless, the entropy concept has been used in several environmental contexts, including extending the MEP to the "environmental and ecological systems" of Earth—and the importance of focusing on how exactly entropy is produced, as pointed out by Volk and Pauluis (2010), or applying the principle to explain the overall self-organizing properties of our Earth—often referred to as GAIA (Lovelock, 1979)—as proposed for instance by Karnani and Annila (2009). Ludovisi et al. (2005) recommend the use of specific dissipation (entropy generation per structure—both expressed in entropy terms) as an indicator.

Entropy in itself is a rather problematic concept because it is not defined in all situations and, in particular, in situations that we would normally consider as representing real-world conditions—systems far from equilibrium. Again, the introduction of the concept of work energy helps in solving this problem, since the definition of work energy as such does not involve entropy.

1.6 Work Energy (Exergy)

Having recognized the previously mentioned problems of resource depletion and pollution, it becomes clear that action needs to be taken and that many of the problems stem from our consumption of various energy or material sources.

Through this recognition, we come to another point, which is that traditional energy budgets following the first law of thermodynamics constitute a proper and practical inventory tool, allowing us to check for energy balance. This, in turn, makes it possible to track what are the big or small flows and the stock in the system, that is to identify items in terms of quantity. However, such an analysis alone cannot be considered sufficient to identify what sustainability is about. First of all, it usually only includes energies, not materials. Second, it distinguishes between the different quality forms of energy, which may turn out to be misleading and guide us in the wrong directions.

One major weakness soon appears, because this type of budget does not tell us whether a big flow is important or not. Is it something that can and shall be reduced? Is it something we can reduce by present or future technologies, or is it simply unavoidable?

To put it in brief, by implementing another type of analysis, namely by converting the previously mentioned stocks and flows into their respective amounts of work energy (aka exergy), we can reveal and identify where we have the most important components. Where do we find significant consumption or severe losses that we should be concerned about? For further definition and an attempted explanation, refer to Chapter 2.

In other words, we attempt to identify the "lowest hanging fruits"—the fruits that are most easily and efficiently harvested. When combining this view with the identification of existing technological possibilities, the cheapest solutions in terms of energy investments or economic costs, we have constructed a management tool that makes it possible to identify appropriate and sensible solutions to environmental problems. This, in turn, should also lead to an increase in the sustainability of our society, be it at the local or regional or, for that matter, global level.

Some examples of the application of the concept at various levels of production and in various processes in our society may be found in the following to illustrate the wide applications that the concept has recently found.

Although the fact that energy could only partially be converted into work was discovered in 1824 by the French engineer Sadi Carnot (1824), the actual importance of its implicit message only seems to have been recognized around the turn of the 19th century. Basically, the core message was that all energy transformations are imperfect; that is they will always lead to a loss in quality. However, different terms were coined, and for a long time in the US, work energy remained in the disguise of an *availability* function, so it was not until the late eighties that the term *exergy* was widely recognized. Here we tend to use the equivalent term *work energy*, as it clearly reflects that our concerns are focused on that part of the energy that is able/available to do work.

While the concept since then has had a profound importance in the engineering world, many of the applications have been dedicated to improving particular processes and/or devices. An overview of types of application may be found in Wall (1977), Gong and Wall (2001, 2016), Wall and Gong (2001) and Vosough et al. (2011).

As the concept allows us to evaluate different types of energy it has been convenient to apply it to assist in the analysis of whole production chains, for example biodiesel production (Talens et al., 2007), cement production (Koroneos et al., 2003, 2005, 2012), the analysis of wastewater treatment plant services (Hellström, 1997; Mora and Oliveira, 2006), solid waste handling (Finnveden et al., 2005; Moberg et al., 2005; Sciubba, 2003; Zhou et al., 2011) and sectors such as food production (food supply chains; Apaiah et al., 2006).

Lately, several cases have been published where work energy has been used to analyze sectors at a national level or even to establish a national budget.

Applying exergy analysis in various forms have, over the recent decades, gained in popularity. The scientific efforts span from various sets of specific activities, such as cement production, through sectorial activities, such as the importance of transports and industry, to the level of countries and even global perspectives. For further references, see Ertesvåg (2001, 2005), Chen and Chen (2009), Wall and Gong (2001) and Gong and Wall (2001, 2016).

In fact, the application of work energy has led to the emergence of a more systematic principle.

Following the emerging disciplines that have been established to observe and estimate the potential impact of products or services on the environment, such as Life Cycle Assessment (LCA) and Material Flow Analysis (MFA), several researchers have attempted to expand these approaches by integrating them with work energy analyses of the systems.

Such analyses have focused on the resource or supply side of the societal systems and have been used as a sustainability indicator for the use of resources (Koroneos et al., 2012). Attempts have also been made to correlate these with outputs or wastes and toxicity (Bastianoni, 1998).

One major advantage, as mentioned earlier, that is expanded on in the following chapters is the fact that work energy serves to combine energetic and material fluxes by bringing the two into a system where they form a common currency. An example of this is the analysis of material and energy use carried out by Ruth (1995). This means that energetic and material fluxes may be compared and weighed against each other, but this does not necessarily mean that they also can be exchanged, that is that a conversion between the two is possible.

Attempts to combine and expand analyses have resulted in the use of Exergy Flow Analysis (ExFA) by Talens et al. (2007) in a material flux-based approach that is close to some of the analyses carried out in this report. Other attempts come closer to the LCA approach in an evaluation of the work energy of mineral resources, carried out as an Exergetic Life Cycle Assessment (ELCA; Mester et al., 2006). In fact, it has been found that exergy gives the more correct estimate of the depletion of resources (Hirs, 2003).

If one accepts the framework described here, it also becomes clear that one may construct a new view of economics, where, for instance, economic evaluations and actions may be more closely based on and related to thermodynamic costs; in other words that irreversible losses of energy also represent some value and hence should receive attention. Such approaches have already emerged and are known as thermo-economics or exergo-economics (Sciubba, 2004; Sciubba and Wall, 2007). All these methods represent new types of energy accounting systems where the First and Second Laws of Thermodynamics are merged but are also combined with an economic valuation of energetic and material losses—the inevitability of which is inherent in the second law.

One major weakness appears when thermodynamic concepts and thus work energy are applied over a wide range of types of systems ranging from classical thermodynamics and ecosystems or even the Earth. The various types of systems are often distinguished by different surroundings or environments which, in turn, leads to a wide range of different reference values to be used in the calculations. In particular, this leads to problems when comparing systems. Thus, for the previously discussed systems (Serova and Brodiansky, 2004), it is important to define the environment more precisely.

This, in turn, also affects the choice of boundaries for the systems under consideration. When other boundaries are chosen, whether scaling up or down, it is likely that the reference conditions will change in accordance with the size of the system. The reference of a system is often chosen as the immediately surrounding environment in which the system is embedded. Attention should be paid to this point during the development of a methodology to describe the work energy situation of larger scale systems.

This influences the estimation of exergy values of the elements or compounds as the calculation and results may vary strongly with the chosen reference conditions (Rivero and Garfias, 2006) and hence the values that may be found for the exergy contents (Szargut et al., 1988; Szargut, 2005).

When focusing on substances, the work energy contents of products may vary greatly over the life cycle of the substances under consideration as demonstrated by the concept of statistical entropy (Rechberger and Graedel, 2002). Generally, during the processing of compounds from their ores and their incorporation into goods, their statistical entropy decreases and correspondingly their work energy density increases until they are disposed of and start their return (dissipation) to the concentration found in Earth's crust, that is following the opposite trend in statistical entropy and work energy.

1.7 Work Energy and Nature

In most of the preceding, we have been dealing with physical or engineering systems as illustrated through the function of factories, plant or individual machines and processes employed in the plant.

In most cases, the estimation of work energy involves a consideration of the energy bound in the chemical components entering and leaving the respective systems.

This fact combines the world of physics and engineering with that of nature, as most of the work energy of biological organisms and ecosystems is in the form of chemical work energy, which, in fact, often can be reduced to a question of Gibbs free energy.

1.7.1 Work Energy of Nature as Chemistry

As the logical outcome of a workshop on biophysics held in the 1970s, S. E. Jørgensen and H. F. Mejer came up with the hypothesis that the function and stability (resilience, buffer capacity) of ecosystems should somehow be reflected in the work energy of the systems (Jørgensen and Mejer, 1977, 1979, 1981; Mejer and Jørgensen, 1979; Jørgensen and Svirezhev, 2004).

If left on their own, the systems would evolve in a manner that would increase their work energy content (density) in accordance with the limitations imposed on the systems and determined by the "prevailing conditions". New species, or, rather, species constellations, able to perform more efficiently and therefore to store more work energy would be selected for and would continuously replace the less efficient constellations, leading to a directional evolution.

This was considered not only to be a "translation" of the "Darwinian dogma of survival of the fittest" into thermodynamics. It also gave us a new way of quantifying the function of ecosystems by knowing their chemical composition. In addition, such an approach was compatible with the bio-geochemical methods often used for the monitoring of ecosystems, methods which developed from the mid-1950s and are still used today. This must be seen as complementary view as compared to the qualitative indicators often used in ecology, such as biodiversity, trophic position and so on.

The calculation of this work energy was based on the chemical composition of the biological organisms as well as the distribution of chemicals among them, which made it simple to apply in the bio-geochemical models of the time.

1.7.2 Work Energy Including Information

Meanwhile, this simple approach turned out to be not without limitations as it led to further considerations. Again, the choice of reference level was important, and what would be a proper level to choose? This problem was shared with many of the industrial systems—but which reference level would be better for nature and environmental systems? Would it be possible to establish one reference level common to all types of systems? These concerns have been continuously addressed over the years, most recently by Jørgensen et al. (2010).

A second issue was that all the biomass in the model of the ecosystem was counted as equal in terms of work energy; that is 1 g of a primitive organism such as single-celled phytoplankton would represent exactly the same value as 1 g of a dolphin. In other words, one gram of a primitive organism energetically counts as the same as 1 g of a much more complex or complicated organism. This may be considered correct from a chemical point of view, since the values we would get when measuring the energy content of samples of biomass from various organisms in a bomb calorimeter will be approximately the same.

On the other hand, the approach seems counter-intuitive, because the dolphin is indeed a much more complex organism than the phytoplankton cells and therefore should contain much more work energy. Or at least it should have taken much more (work) energy to establish the more advanced organism. The answer seems relatively simple: there is work energy contained in the information needed to construct them, and this makes the difference.

Therefore, it was proposed that the chemical value of the biomass of the various biological components of the ecosystem should be weighted in order to reflect the amount of information used to construct them (Jørgensen et al., 1995; Jørgensen, 2006). Since then, much effort has been invested in deriving the weighting factors for various organisms derived from the size of the respective genomes (Fonseca et al., 2000; Jørgensen et al., 2005; Ludovisi, 2009; Ludovisi and Jørgensen, 2009).

In brief, this new work energy which differs in its foundations from classical work energy or exergy is nowadays referred to as "eco-exergy" or "work energy including information". For a more thorough discussion of this theory and related thermodynamic theories of ecosystems, the reader is referred to Jørgensen (2002) and Jørgensen and Svirezhev (2004).

Thus, a new method has been developed to estimate the additional effect that the intrinsic information of organisms has on the "value" of nature, and it would seem obvious to include such considerations in an (e)valuation of society—especially because we today talk of our social constructs and the present "digital age" as the "information society". Surprisingly, no attempts have been found in the literature to quantify the importance of this part of our society to our own reproduction—as, for instance, compared to the energetic costs of this investment. Therefore, it was chosen to base the current comparison and its conclusions entirely on the physico-chemical values of structures.

2

Work Energy and Sustainability

2.1 Introduction to Work Energy

In this chapter, we continue the introduction to the previously discussed new way of viewing energy, with its stocks and flows, in general, and of applying this view to our society. In fact, we introduce several new perspectives that will help us to get closer to the crux of sustainability, in other words—hopefully— to understand better what sustainability means and how we may achieve it.

Meanwhile—as it might also be concluded from Section 1.6—to claim that these concepts and perspectives are completely new is a little erroneous. So, when the newness of the approach is stressed, this does not mean that everything is new: much of the knowledge used here existed already. The really new aspect is rather the way we are using this knowledge. We just have to take everything that we more or less already know implicitly and put it together in a different way. In a sense, the approach is quite obvious, and it is rather strange that this way of putting things together has not become more popular and has not been done more frequently hitherto.

In fact, for the perspectives presented here, when seen as consisting of the partial compositional elements of a theoretical framework, we may find that there is very little new in the individual parts. For example, we know that heat is lost in almost any process that we make use of in our society to support our everyday existence. So the new component may eventually turn out to be the way the various elements of already-existing knowledge are put together and used. That is the merger of existing theories and practical knowledge and experience, when assembled in a new manner, provides us with results which give a broader picture of the whole system. However, such a merger has to be carried out in several steps.

Going back in time, ever since the work of the French engineer Sadi Carnot (1796–1832) it has been known that a perfect (100%) conversion of energy into work is not possible; that is a perpetual motion machine is not possible (Carnot, 1824). During the transformation of a given amount of energy a part of it will always be lost due to the "friction" of the system. This part simply "disappears" (dissipates) and is no longer available to us. Such a simple semantic exercise combines the initial basic observations about losses during energy conversion in heat engines with the later works of Gibbs, Prigogine and co-workers.

It was these early, preliminary observations of Carnot that led Rudolf Clausius (1822–1888) to the formulation of the Second Law of Thermodynamics around 1865 (Clausius, 1865). This was actually before the first thermodynamic law had been formulated. So most confusingly, of the thermodynamic laws we use here (see Section 2.4), the second law was the first to be discovered and formulated.

One fundamental problem exists and is likely to remain unsolved years from now. The second law of classical thermodynamics is held by many physicists to be valid for (ideal) gases only. Furthermore, the law is viewed as applicable only under conditions where the system is very close to thermodynamic equilibrium. Clearly, we have a conflict here, since real systems are composed of many other states than ideal gases only, and thermodynamic equilibrium represents a state that is never present anywhere where life is found, hence also the discussion relating to the lack of a definition for entropy. When applying thermodynamic theory or adopting a thermodynamic approach to life and the circumstances pertaining to it, we need to realize that life is not a gas as postulated by Skene (2015) but at the least is something apart and very different from gases. Thermodynamics has here been transferred to another domain where most of the general equations—although formulated in a mathematically speaking isomorphic manner—simply have lost their meaning in a traditional sense and must be re-interpreted to give them another meaning (cf. Deacon, 2007, 2008).

If one insists on retaining the strict physical view of thermodynamics, much of modern science and many hardcore disciplines, such as physical chemistry or chemical engineering, will be heavily constrained in their practices, to say nothing of the fact that we will have to abandon our hopes of expanding thermodynamics to biological or even societal systems. Such an initiative does not make sense. Luckily, we also find physicists that are supportive of this expansion of thermodynamic views to other conditions or domains.

Erwin Schrödinger (1887–1961) was probably one of the first to apply a strictly thermodynamic view to systems belonging to the biological domain, although as early as 1922, Lotka (Lotka, 1922a, 1922b) made the first statements about organisms maximizing power and life being a struggle for (Gibbs) free energy. Essential in this context is Schrödinger's famous—but also rather unfortunately formulated—statement that living systems should be "feeding on negentropy" (Schrödinger, 1944). This leaves the impression that entropy can be negative, a statement that is strongly in conflict with the normal formulation and also a contradiction of the Second Law—namely that entropy is always positive having zero as its lowest value. The conflict is semantic in character and must be regarded as an artefact derived from an imprecise or thoughtless formulation (see beginning of Chapter 3). In addition, this must also be seen as an unfortunate consequence of moving between domains—and a view of entropy as describing states or belonging to processes that bring systems from one state to another.

Thanks to the work of the Russian/Belgian thermodynamicist Ilya Prigogine (1917–2003) and his colleagues, some progress has been made in this area (see for instance Glansdorff and Prigogine, 1971; Nicolis and Prigogine, 1977). Building on the works of the Norwegian physicist Lars Onsager (1903–1976) the thermodynamic approach was extended to apply to states that are far from equilibrium—and thus to conditions valid to most real-world systems. It turns out that with the same assumptions systems that are far from equilibrium tend to develop towards a state of minimum energy dissipation (density). This finding is in contradiction with statements by the many physicists who work with the maximum entropy production principle of Swenson (1989) and Jaynes (1957a, 1957b), and the debate around this remains to be resolved (Aoki, 2018).

In this manner the views of Erwin Schrödinger (formulated and published in 1943–1944) could be refined: as Prigogine pointed out, systems were exploiting the energy gradient imposed on them, using the energy released by irreversible processes (see Brillouin, discussed later) to build up structures that broke down the energy even further—but with decreasing entropy formation per unit, so-called dissipative structures.

What is left now is to answer the question of what exactly such structures are doing? Do they try to conserve energy? Or do they break it down as fast as possible? Do they minimize or maximize their dissipation? Is it really a minimization or a maximization or something different like adapting their development to make the best possible use of the energy constraints on the system? In their use of energy do the systems optimize their processes? To answer these questions is far beyond the scope of this report, but it should be kept in mind that these questions are reflected in many papers of today. Unfortunately, in many cases, the difference in viewpoints serves only to blur the scientific debate, amongst other things disturbing discussions about issues such as what sustainability is about.

From the beginning of the 20th century, we may identify "hot spots" where interest in the message from the second law—the imperfect conversion of energy and the fact that part of it was inevitably transformed and lost—led to an increasing awareness of the fact that only a certain part of the energy is available to us. Several terms have been applied to this part, such as *essergie*, *arbeitsfähigkeit*, *availability* and *exergy*. As all these terms deal with the same aspect, namely the capacity of a given amount of energy to do work, we will here use *work energy* to designate this capacity and use it in a manner that makes it synonymous and exchangeable with *exergy*.

2.2 Introducing Energy Forms to Everyday Life

The most fundamental thing about work energy is the need to see energy not just as energy but as energy in various forms and capabilities of doing work.

This first point of shifting forms might seem obvious, but what is less obvious is that the different forms of energy are not equally good at doing work. This is what the work energy (exergy) concept is about. Thus, work energy adds a quality aspect to our view.

In the way we talk and think about energy in our everyday life, we may implicitly know most of this already, but often it seems that this knowledge has not taken root in our minds—at least not enough to be put into practice. In fact, the implicit aim of initiating a project like the one presented here was to shed new light on the role of the various activities we are undertaking and to lay bare all (ir)rationalities.

During the energy crisis in the early 1970s, we became aware of the important role played by oil and coal as the primary energy carriers used to run our societies. The consumption of these fossil fuels supplied us with electricity and heat as well as the gasoline and diesel for transport. Nuclear power plants were set up, as they were seen as an efficient or necessary alternative to fossil fuels, but the dangers connected with this type of energy production became clear following the accidents at Three Mile Island (1979), Chernobyl (1986) and, more recently, Fukushima (2011), not to mention the ongoing discussions about what to do with the radioactive waste products when nuclear power plants are to be decommissioned.

Since then we have had an ongoing discussion concerning possible alternatives to these finite (non-renewable) energy resources, focusing on other types of energy based on renewable energy resources such as photovoltaics, wind power, hydropower and perhaps biofuels. Most of us will also know that we need energy to sustain our lives and that different food items contain different amounts of energy in their constituent sugars, fats and proteins. We also know that systems using energy get warm: electronic equipment around our house gets warm when in use; doing exercises, we start to sweat to get rid of excessive heat released by the combustion processes in our body.

In fact, having reviewed this relatively simple knowledge we have revealed most of the basic forms in which energy exists. The electromagnetic forms such as atomic or solar radiation (e.g. nuclear power and photovoltaics), kinetic and potential energy (wind and hydraulic power), chemical exergies (such as the energy found in food and material) and heat. Heat is most often seen as the end station of this chain as it represents almost no work energy. However, this does not mean that energy in this form should always be ignored. For instance, in some recent technologies, heat at high temperature may be stored and later used to make steam to power electricity production or, at lower temperatures, heat exchangers may be an important component for deriving energy from (bio-geo)chemical processes.

We also implicitly know that we can convert certain energy forms into others. Kinetic and potential energy can be turned into electricity by wind and hydropower installations, the chemical energy of oil can be used to heat a house, the chemical energy of gasoline can be converted into kinetic energy when running a car and so on. Figure 2.1 demonstrates a number of the conversions between different energy forms that occur in our everyday life.

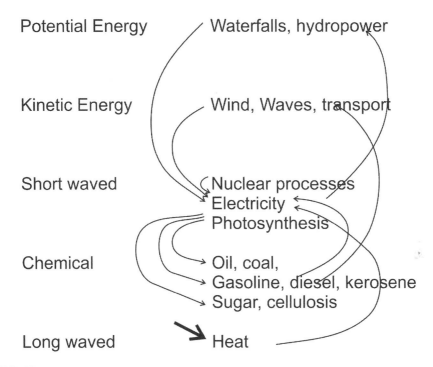

FIGURE 2.1
A diagram showing some of the transformations between various forms of energy that occur in our everyday life. The diagram only considers that the conversion may take place, not how it is possible, that is what technology is needed to perform the operation.

Meanwhile, it is important to notice that every time we convert one form of energy into another, there is a cost in terms of the dissipation of the energy needed to make the conversion possible; this cost ultimately ends up in universal entropy.

In conclusion, we seem to possess a lot of knowledge about work energy. All we need is to organize this knowledge in a more systematic way for it to be useful. While a diagram such as Figure 2.1 shows that the conversions are possible it does not tell the full story. How does the conversion happen? How is it made possible, and how easily is it achieved? What are the costs— in terms of both energy and economics? Do we have the necessary technologies available to us?—and so on.

2.3 The Relevant Law(s)

The First and the Second Laws of Thermodynamics are both essential to the analysis of the energy supply systems of today. However, analyses based on the first law are still far more common at the macroscopic level (demonstrated, for instance, by national energy accounting systems and energy budgets in many countries), whereas analyses based on the second law have mainly been implemented at lower levels for instance when optimizing factory equipment, production processes or assembly lines.

There is no doubt that first law analysis provides us with a brilliant tool when serving purely as an accounting system to keep track of energy flows. But as mentioned earlier (Section 2.2), the first law does not distinguish between the various types of energy. To the first law, all amounts of energy that are equal are the same—1 kJ of energy in the form of solar radiation is the same as the energy contained in 1 kJ of sugar or 1 kJ of heat at 20 °C. The first law simply ignores the form or type of energy and its inherent capacity for doing work (quality). Energy is energy is energy, to paraphrase Gertrude Stein.

This understanding gives a misleading picture, because the three types of energy mentioned earlier differ from each other in one important aspect—namely their ability to do work—which, in the three instances mentioned previously, is quite different. So, if we want to do work it does matter greatly to us—not only that we have a sufficient quantity of energy but also in what form we have the energy, that is that we have the necessary quality of energy. So, when it comes to doing work, the value of each of the three amounts of energy—the arbitrarily chosen 1 kJ—is not the same. This is expressed by the work energy content, the availability of the energy to us or the exergy.

Implicitly this means that out of a given amount of energy we have a part that may perform work and in addition a part that is not able to do work (anergy—the Greek word for being unemployed is *anergos*, "not having or doing work"), leading to the following equation:

$$\text{Energy} = \text{Exergy} + \text{Anergy}.$$

In order to incorporate this view—the different capacity of various types of energy to do work—we need a new framework to carry out these analyses (Nielsen and Bastianoni, 2007).

This is where the second law and work energy analysis come into play. Work energy reflects a quality possessed by the different forms of energy and is also able to quantify it. The different energy forms—even when equal in quantity—possess different abilities of doing work. This is basically what work energy expresses. Two equal amounts can be found where the one has a high ability to do work (high work energy content, e.g. radiation) and the other has a very low ability (low work energy, e.g. heat at 60 °C). Examples of energies with varying contents of work energy are shown in Figure 2.2. This way of carrying out analysis is often referred to as thermo-economics, exergonomics and several other similar terms.

Generally, in our efforts to optimize energy systems, we should always concentrate on the amounts of energy that possess a great capacity to do

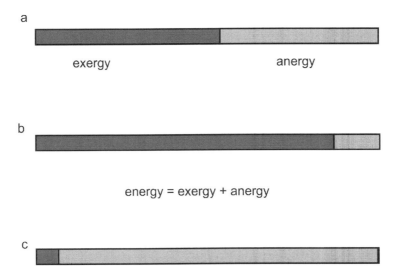

FIGURE 2.2
Equivalent amounts of energy split into their respective contents of work energy—the part of the energy able to do work (dark, exergy) and the part that is not able to do energy, that is cannot be used—the anergy part (grey). Part (a) shows the split and an intermediate situation, (b) shows energy with a relatively high content of work energy and (c) the opposite, a low work-energy content.

work. For instance, both inputs and outputs might be unnecessarily high indicating inefficient systems. This means that by focusing on work energy we almost immediately get a message about where to concentrate our efforts in order to improve the performance of the system under observation. This forms part of what is referred to when talking of "picking the lowest hanging fruits".

The only question left is if there is any general pattern in the contents of work energy in different energy forms. Are the conversions between them following any rules that we need to observe? And is there, in fact, a system that should be observed and complied with when measures need to be taken?

2.4 Work Energy Content and Transformation Rules

It was the French physicist Leon Brillouin (1889–1969) who was one of the first to observe an overall system in this. He pointed out, first, that some forms of energy had a higher work energy (as introduced previously) and, second, that there was a hierarchical order amongst them (Brillouin, 1962, originally 1949). Third, he found that there was also a directionality in the transformation, since spontaneous transformations would always happen from a higher to a lower content of work energy. This sequential ordering is an implicit consequence of the second law.

Following the preceding, at the high-quality end of the range, we find the radiative and electromagnetic forms of energy such as nuclear power and electricity, which have a high content (100%) of work energy. At the low-quality end, we find energy in the form of heat, and as we approach heat at ambient temperature, we also approach a work energy content of zero (0!). At intermediate levels, we find all the chemically bound energies, some of which—like fossil fuels—have a high work-energy content and others which have much lower values, for instance biomass and small molecules. At the chemical level, the energy densities may vary widely. This ordering may be illustrated in Figure 2.3.

In general, there is a continuous transformation from higher to lower level. As previously mentioned, going upwards in this hierarchy is only possible at the expense of an additional input of work energy, and the cost will be raised by the breakdown of these additional inputs, that is the loss in quality (dissipation) of the extra work energy.

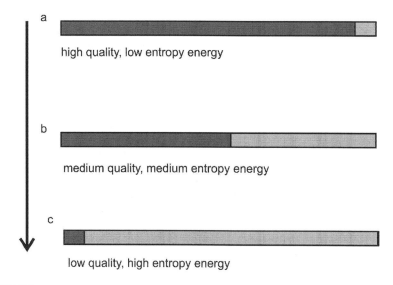

FIGURE 2.3

The conversion of a given quantity of energy can only take place in one direction, from a higher to a lower content of work energy—part (a) may now symbolize radiative, electromagnetic or high-density chemical energies with a conversion factor between energy and work energy close to 1; (b) illustrates the chemical energies with intermediate work-energy values; and (c) the thermal energies with low work energy contents approaching zero when getting equal to ambient temperature.

The difference in work capacity between similar amounts of energy as demonstrated by Figure 2.3 also means that we are now able to convert a given amount of energy to its work energy content by multiplication by a factor (ξi) expressing the relative work energy content of the respective type (i) of energy. The work energy (WE) of type i is thus

$$WE_i = En_i \cdot \xi_i.$$

Some typical values used for the conversion of energy into work energy (exergy) content are shown in Table 2.1.

Taking equal amounts as an example we can now concentrate on the work energy contents. As energy is always conserved, energy efficiency will always be 100%—since nothing has disappeared—but the work energy efficiency is something quite different.

So, to summarize, we have now divided

1. energy in different forms into
2. different contents (fractional parts) of work energy.

The problem we will face in the following is that having a high energy content does not necessarily mean that this work energy can also be used (e.g. the example of nuclear power given in Table 2.1).

We are now able to identify the largest consumptions of work energy either in terms of inputs/imports or as the largest losses—and we can do this for energy conversions occurring in processes going on in a factory or in our society. Both will be relevant when attempting to optimize such a system,

TABLE 2.1

The Differences in the Quality of Different Forms of Energy Expressed by Relative Contents of Work Energy (WE) Together with Examples of Types, Values and Use

Level	Form	Approximate WE Content (Fraction)
High-quality energy	Nuclear power	1
	Solar radiation	1
	Electricity	1
Intermediate	Oil	1
	Coal	0.9
	Sugar	?
Low	Heat > 100 °C	0.6
	Heat < 100 °C above ambient temp	0.2–0.3
	Heat at ambient temp	0

Source: Values are taken from several sources: Sciubba and Wall (2007), Dincer and Rosen (2007), Szargut et al. (1988) and Szargut (2005).

that is to achieve a more efficient use of resources. This goes for both energy and matter.

2.5 Using Work Energy to Give Priorities

In the previous section, we have established a framework for distinguishing between energy and its work energy content and for understanding the relatively simple pattern underlying the conversion rules. But how can we use this knowledge to improve conditions and optimize a system?

We can start by splitting Figure 2.2 into yet another fraction. In principle, the anergy part is uninteresting to us, but we will keep this part of the illustration to maintain the connection to the first law. So in the framework to be established, the same strategy will be followed, since it is important to track both types of losses—the losses that cannot be avoided and the losses that can be avoided or that still contain useful amounts of work energy.

In short, the only interesting part of the energy is the part that is able do work (i.e. work energy, exergy). So, as a second step, we will identify how much of this that is actually used. In other words, we accept the possibility that there is a fraction of the work energy that is not exploited by a certain process—that the whole potential is simply not used. The reasons for this may be many. Mostly it is due to limitations in the technologies currently available to us. This is the area where many applications of exergy in the engineering sciences are found, but it seems that for any specific technology, in many cases a maximum degree of exploitation exists. In such cases, any real improvement of the situation requires the development of a totally new technology.

Various situations showing hypothetical outcomes of an analysis are illustrated in Figure 2.4.

In the upper part of Figure 2.4, we identify both the dark and light grey parts of the diagram as the total work energy capacity of the given amount of energy. The dark part symbolizes what is actually used, while the light grey part in principle is available to us but, in fact, is left unused for reasons such as the ones mentioned earlier. The parts are almost equal, leaving one third for *anergy* and two thirds for work energy, but only half of the work energy is exploited by the process under consideration (Figure 2.4, part a). This signals to us that the yellow part shows a potential for the efficiency of the process to be improved or, alternatively, that the work energy could be transferred to other processes. Should we regard this amount to be "lost" in our accounts, meaning that we have also lost energy of high potential value, that is potential work energy? If easily implementable technical solutions are already known to exist, this would be a typical "low-hanging fruit".

The situations in (b) and (c) show us the situation for an amount of energy which, in both cases, possesses a high work-energy content. Only in situation

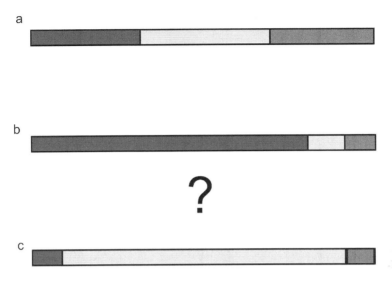

FIGURE 2.4
The work energy (WE) part for three equal amounts of energy is shown. Part (a) shows the classical view where the part of WE which is not exploited is grey, so total WE equals the sum of the dark and light grey areas (grey still represents anergy). In part (b), we find energy with a relatively high amount of WE and only a small unused part, that is a high-WE efficiency, whereas in part (c) for the same amount of WE, we find only a minor part used and a large proportion unused, that is a low-WE efficiency.

(b) is most of the work energy exploited. This means that we have a high efficiency with respect to work energy. A further increase in efficiency will rely on further optimization of the system, which might not be possible. After all, the second law tells us that also in this case reaching 100% efficiency will not be possible.

In situation (c), most of the work energy (capacity) is left unexploited, and the system has a low work-energy efficiency, signalling that we should make efforts to carry out this process much better, that is to increase the efficiency. In some cases, this is possible with only slight effort because we can easily identify an existing technology to assist us in solving the problem. If the technology is already well established, it may even be cheap to implement, as is often the case within the discipline of cleaner production. Many examples of this type exist nowadays, where, for instance, pumps and heating devices are getting more and more efficient. As mentioned, in other cases, the technology does not yet exist, and a solution will have to be based on developing such new technology.

Another issue will be the economic aspect, as just mentioned. During an analysis, we may identify many situations like the one depicted in situation (c), where a variety of solutions are offered. Usually, everything else being equal, it will be logical to choose the more cost efficient of the solutions.

2.6 Sub-Conclusions Work Energy and Sustainability

- We need to establish a technique/tool/procedure that allows us to combine and compare the energy costs of activities going on both in our society and in nature. Thermodynamics and, in particular, the concept of work energy seem to provide us with such a tool.

- Both types of systems are domains quite different from that of classical thermodynamics and as such are not covered by traditional thermodynamics. Thus, traditional concepts and principles cannot be considered as homologues, and much work is needed on the specifications of the transferred concepts.

- Both systems differ, for example in their reference levels, and for a comparison to be established, we must identify a reference level that satisfies both systems

- The new approach, taking its point of departure in the Second Law of Thermodynamics, contains some inherent messages for us to consider: all forms of energy are not equally good in producing work, energy has a quality side to it and this energy quality is expressed through its ability to do work.

- Our society, in particular, is constructed around high-quality forms of energy (found as radiation and electricity), intermediate forms are chemical-bound energies and low-value forms include heat.

- Spontaneous change or transformation of energy occurs in one direction only, from higher to lower forms.

- In our construction of society and use of technologies, this ability to do work is not always exploited fully, and this unexploited ability to do work is an obvious target for investigation and for development of measures to improve the situation.

3

Methodological Considerations

3.1 Introduction to Methodology

This chapter presents further considerations about the various proposals introduced in the preceding chapters, namely how to integrate them with society. This includes the concept of sustainability, the relationship between renewable vs. non-renewable resources and how to estimate the respective work energy contents of either the energy or the materials that we use to power the activities we undertake. Such a merging of concepts represents an attempt to establish an evaluation system that will serve not only to define sustainability but also to assist in guiding our society in the direction of continuously increasing sustainability, thus facilitating its transformation into a genuinely sustainable society. Optimally this will happen in a systemic (i.e. taking society and all its components as an interdependent whole) and systematic manner. This means that the approaches that are developed and applied should be generally valid, that is generic, in the sense that it should be possible to apply them to the analysis of sustainability at all sociological and geographical scales—from small areas to more extensive regions and eventually to society as a whole.

Thus, two basic ideas and assumptions lie behind the approach developed here:

1. All activities in society, including the exploitation of resources, and infrastructure (both administrative and domestic), are in some way correlated with a geographical location. Infrastructure and societal activities can thus be correlated to information about the geographical, spatial organization of our society.

2. In order to identify the quantitatively important parts of our society and be able to address sustainability in terms of work energy, it is convenient—if not completely essential—to make a certain degree of deconstruction of our society into activities, although a trade-off has to be established, since both too few and too many details may result in the outcome of an analysis becoming too obscure, the whole picture being blurred or even inherently contradictory, making it impossible to identify any measures to be taken.

As a consequence,

1. the "landscape" (in a broad sense) of the society under consideration needs to be divided into its geographical compartments, covering everything from the extraction of resources, production activities, the location of its infrastructure, countryside and so on. The first step is therefore to divide our society into geographical sectors. Here the methodology of the Corine Land Cover (CLC) system—commonly used in the European Union—has been adopted, as described in Section 3.2. This must be seen as a pragmatic solution ensuring the compatibility of the approach taken here with similar studies in other countries in the European Union. Similar data systems are likely to be found today in most countries. Meanwhile, in many cases the data sets available in the official databases are not spatially detailed enough to permit accounting estimates to be made for small regions with reasonable accuracy.

Next, referring again to the statements made earlier,

2. it is clear that the societal activities implicitly reflected in the CLC system need further subdivision and analysis, taking into account their relative importance in the societal system under evaluation, in order to be able to derive any possible conclusions about specific sustainability issues.

The most important factors have been found to be

- identification of the necessary societal infrastructure and its existence (see Chapters 4 and 5) and its procurement—both as administration and maintenance, that is all that is supporting the citizens;
- activities establishing and sustaining the everyday life of citizens in private households (Chapter 6), that is the everyday life of the individual;
- agricultural activities, that is crops, livestock and food production—and any related exploitation of natural resources (Chapter 7);
- industrial activities, that is production per se, and trade and commerce (Chapter 8);
- the extent of non-exploited nature, supplying society with unpriced benefits such as ecosystem services (Chapter 9); and
- unexploited resources in terms of wastes (Chapter 10) and surplus material from nature (from Chapter 7).

It will be seen that this is basically the scheme followed in this report. A condensed version of the perceived structure is given in Figure 3.1.

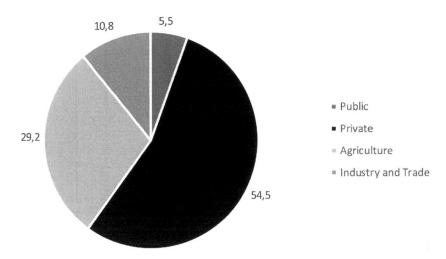

FIGURE 3.1
The partitioning of infrastructure between the four major sectors involving material stocks
(units: %)

Combining this with the approach presented in the previous chapters, it means that we are able to estimate and monitor the necessary structures, quantifying the inputs and the outputs either as the inevitable losses, as well as the still useful energies exported, all accounted for in terms of work energy. At the same time, the structures and activities can be related to their extension (either in terms of geography or their actual size) and combined with traditional energy and material flows to form a consistent basis for an evaluation of sustainability issues. The link to an economic valuation will be left out here as it is outside the scope of the project. However, one should not forget that the economy is an important and powerful tool in the regulation and governing of activities of the system.

From a historical point of view, this method is indebted to the works of Prigogine and his co-workers (Prigogine and Stengers, 1984), who introduced the idea of physical and biological systems as dissipative structures. This view was later expanded to encompass society and economy as presented by Georgescu-Roegen and Rifkin and others who introduced the concept of entropy into our views of society (see Georgescu-Roegen, 1971; Rifkin, 1989).

At the same time, but in a parallel development, second law analysis— including investigations on work power (exergy) balances—has become a normal starting point for the optimization of industrial systems, in particular during the 1990s (see Cornelissen and Hirs, 2002; Dincer, 2002; Hammond, 2007; Bastianoni and Marchettini, 1997; Svensson et al., 2006; Sciubba and Wall, 2007; Dewulf et al., 2008; Chen and Chen, 2009; Valero, Valero et al., 2010)

Although the idea of merging the two directions seems obvious, it has taken some time to develop and, in particular, to set up a convenient framework for such an analysis (Nielsen and Bastianoni, 2007; Bastianoni et al., 2006). It has become increasingly necessary to apply such methods in order to guide a society or a geographical region in the direction of sustainability or ultimately to achieve true sustainability (Valero, Uson et al., 2010).

More recently, the ideas have been refined and proposed to be a possible way to integrate and evaluate the cycles of energy and matter flow that occur within our societies as well as in nature. The idea that society could be understood as an ecosystem and that studies along the lines given in modern ecosystem theory (Jørgensen, 1992; Nielsen et al., 2020) should be used to understand and re-organize societal activities in a way that mimics the development and function of ecosystems was argued by Nielsen and Müller (Nielsen, 2007; Nielsen and Müller, 2009).

In order to achieve an understanding of the use of energy and matter in society, one obvious approach is to follow the framework laid out in guidelines for Life Cycle Assessments (LCA; Bauman and Tillman, 2004) combined with Material (and Substance) Flux Analysis (MFA or SFA; see for instance Baccini and Brunner, 2012; Hannon et al., 1993).

Although both methods have become standardized over the past decades, both LCA (Bauman and Tillman, 2004) and MFA/SFA (Baccini and Brunner, 2012) still have some basic inherent problems that—amongst other things— make comparisons between even quite similar functional products quite cumbersome and problematic. This arises from issues such as where to define the boundaries of the system. From LCA, the concept of *cradle to grave* emerged, so, in fact, we need consistent ways of defining where is the cradle and where is the grave. Another metaphor cradle to cradle appeared in connection with the many ideas launched together with the Circular Economy concept, where the possibilities of the many Rs became apparent (Reduce, Reuse, Recycle, Repair, etc.; Mester et al., 2006, 2009)

As the next step, it has been proposed to integrate work energy, aka exergy, with many of the previously mentioned approaches for analyzing the life cycle of products, an approach that is normally referred to as Exergetic Life Cycle Assessment (ELCA; Cornelissen and Hirs, 2002).

Using work energy or exergy as the basic unit for the analyses makes it immediately possible to compare the invested energy and matter flows in activities, products, goods and services, since they can be converted and recalculated so that they share the same units as energy (e.g. kJ, kWh).

When combining work energy and LCA it has become a logical development to involve also the work energy used in the fabrication of a specific material product or chemical compound, thus leading to the cumulative work energy or exergy value (cumulative indices). This, in other words, takes the *hinterland* into account and expresses the *ecological backpack* of a given product, good or service.

In a few instances, in this study it has, in fact, been necessary to use cumulative values since real exergy values could not be identified. This is possibly due to the fact that the calculation of such values, for instance by following the schemes laid out in Szargut (Szargut et al., 1988; Szargut, 2005), is far from trivial and takes quite a large amount of chemical knowledge to implement, for example in all cases of fertilizers and chemicals (for a recent exergy calculator, refer to www.exergoecology.com/excalc).

The double-dual character of the work energy flows as mentioned earlier is taken as the entry point of this analysis and used to organize and structure the knowledge gathered. Thus, flows will be subdivided into what are believed to be the sustainable and non-sustainable portions as well as being characterized as either energy or matter flows; see Figure 2.2.

In accordance with the framework indicated earlier, the activities have, in turn, been divided in their respective inflows and outflows which also must be compared to their respective resulting stocks (see Sections 3.6 and 3.7) so as to answer questions such as what is the actual efficiency? What are the relative expenses of maintaining a certain activity or structure?

It is the core idea behind this "exercise" that in this manner, it will be possible to establish various efficiency measures based on a somewhat traditional but also expanded I/O analysis, such as Output/Input, Stock/Input, Output/Stock and so on, all accounted for in terms of work energies. It is also possible to divide the flows into renewable and non-renewable parts and thereby make a clear linkage to sustainability issues.

3.2 Basic Information on the Target of Study, Samsø

Samsø is the 13th-largest island in Denmark in terms of area. The official area is 114,26 km², and in 2011, the number of inhabitants was 3,885 but has slowly decreased to about 3,700 according to Statistics Denmark (Danmarks Statistik, 2011).

The order of magnitude of the area also illustrates that the resolution of the CLC data from 2006 is much too low to achieve a reasonable certainty for determining various activities in the area. It has therefore been decided to extract data from other sources, that is mainly from the Geographical Information System (GIS) information inherent in the maps provided by the databases of the municipality which are used for administrative purposes.

The following outline of a methodology has been divided into four sections. The first section deals with the relations between the island and the landscape types in general based on the CLC system (EEA, 2007). The second is concerned with the estimation of infrastructure and its related work energy. The third part is concerned with the work energies of various materials and chemicals that are involved in societal activities. Finally, the fourth part outlines how it is possible to put this knowledge together and—if possible—impose it on the sectorial view for a more thorough analysis of a full society.

3.2.1 Samsø and the CLC System

According to Stjernholm (2009), only 30 out of the 44 CLC types are present in Denmark and thus relevant for this study. A minor subset of these types is found on Samsø and may from time to time be determined subjectively. This is not optimal, but it is not always possible to extract the information directly from the GIS-data which is gathered strictly in accordance with the CLC codes.

This means that the presence of some area types has been identified on a subjective level, for instance by visual inspection, but that an actual and precise quantification of the area or component elements has not been possible (as indicated by nd [not detected?] in Table 3.1). In most cases, the area overlaps others and thus has incorrectly been included in other components. However, the errors arising from these problems are considered to be of minor importance.

An attempt to check how much of the island area is accounted for when following this method of aggregating data—taking into account the uncertainties introduced by the previously mentioned possible overlaps—is presented in the following and summarized in Table 3.2.

The section dealing with terrestrial nature (see Chapter 9) requires the boundaries between the different types of terrestrial ecosystems to be determined. There is a relatively high degree of uncertainty in these determinations, for instance in the estimation of the areas of verges and shoulders adjacent to the various types of roads. Nevertheless, such boundaries are important zones of the landscape that may exhibit "extreme" activity—so-called ecotones—and as such, they may be very important in various ways in the transformation, breakdown or transfer between systems of many biogeochemically important compounds.

It is likely that much of the discrepancy between the accumulated data and the actual size of the island may arise from the uncertainties in the determinations of marginal areas. This may affect assessments of the potential role of such transition zones.

The area of the adjacent estuaries, coastal and open waters around Samsø, has been determined to be approximately 1,438 km^2. This area, corresponding to approximately 12 times that of the island, has been excluded from this investigation as its role in the social sustainability of the island is considered to be very low.

TABLE 3.1

The 30 CLC Code Types Present in Denmark Together with an Identification of Their Presence (p) on Samsø, as well as an Indication on Their Determination (d) or the Opposite (nd)

Level 1	Level 2	Level 3	Present/ Distinguishable	Sector/ Chapter	Remarks
1 Artificial areas	11 Urban fabric	111 Continuous urban fabric	p/nd	5, 6	
		112 Discontinuous urban fabric			
	12 Industrial, commercial and transport units	121 Industrial and commercial units	p/d	8	
		122 Road and rail networks and associated land	p/d	5	
		123 Port areas	p/nd	(5)	
		124 Airports		(5)	
	13 Mine, dump and construction sites	131 Mineral extraction sites	p/nd	(8)	Sand and gravel
		132 Dump sites	p/nd	(5)	
		133 Construction sites			
	14 Artificial, non-agricultural vegetated areas	141 Green urban areas		-	
		142 Sport and leisure facilities			
2 Agricultural areas	21 Arable land	211 Non-irrigated arable land	p	7	separated in crops
	22 Permanent crops	222 Fruit trees and berry plantations	p	7	do
	23 Pastures	231 Pastures		7	do
	24 Heterogenous agricultural areas	242 Complex cultivation patterns	p	7	do
		243 Land principally occupied by agriculture with significant areas of natural vegetation	p	7	
3 Forest and semi-natural areas	31 Forests	311 Broad-leaved forest	p/nd	9	Aggregated data
		312 Coniferous forest	p/nd	9	
		313 Mixed forest	p/nd	9	
	32 Shrubs and / or herbaceous associations	321 Natural grasslands	p/d	9	
		322 Moors and heathland	p/d	9	Overlap possible
		324 Transitional woodland / shrub	p/d	9	

(Continued)

TABLE 3.1 (CONTINUED)

The 30 CLC Code Types Present in Denmark Together with an Identification of Their Presence (p) on Samsø, as well as an Indication on Their Determination (d) or the Opposite (nd)

Level 1	Level 2	Level 3	Present/ Distinguishable p/nd	Sector/ Chapter	Remarks
	33 Open spaces with little or no vegetation	331 Beaches, dunes, and sand plains		9	Limited importance
4 Wetlands	41 Inland wetlands	411 Inland marshes		9	
		412 Peat bogs		9	
	42 Coastal wetlands	421 Salt marshes		9	
		423 Intertidal flats		9	
5 Water bodies	51 Inland waters	512 Water bodies		9	
	52 Marine waters	521 Coastal lagoons		9	Excluded
		523 Sea and ocean		9	Excluded

Note: The indication in column sector/chapter serves as an indicator of where to find more detail information on the calculation(s). p = present, nd = not distinguishable/determined in detail; code in sector referring to chapters: 5 public, 6 private, 7 agriculture, 8 industry, 9 nature (parentheses also refer to pnd).

TABLE 3.2

The Landscape of Samsø Divided according to the CLC System Together with the Respective Areas Estimated from Various Sources

Level 1	Level 2	Level 3	Type	Area (km²)	Remarks
1 Artificial areas	11 Urban fabric	111 Continuous urban fabric		1.63	City
		112 Discontinuous urban fabric		3.24	Countrysides
	12 Industrial, commercial and transport units	121 Industrial and commercial units			
		122 Road and rail networks and associated land	Roads I	2.000	Paved roads
			Roads II	0.442	Non-paved
		123 Port areas	p/nd	—	
		124 Airports	p/nd	—	
	13 Mine, dump and construction sites	131 Mineral extraction sites	p/nd	0.073	Raw materials
		132 Dump sites	p/nd		
		133 Construction sites			
	14 Artificial, non-agricultural vegetated areas	141 Green urban areas	p/nd		
		142 Sport and leisure facilities			
2 Agricultural areas	21 Arable land	211 Non-irrigated arable land		83.26	Not separated here
	22 Permanent crops	222 Fruit trees and berry plantations			
	23 Pastures	231 Pastures			
	24 Heterogenous agricultural areas	242 Complex cultivation patterns			
		243 Land principally occupied by agriculture with significant areas of natural vegetation			
3 Forest and semi-natural areas	31 Forests	311 Broad-leaved forest	Mixed, deciduous evergreens	8.812	Not distinguished in tables
		312 Coniferous forest			
		313 Mixed forest			

(Continued)

TABLE 3.2 (CONTINUED)

The Landscape of Samsø Divided according to the CLC System Together with the Respective Areas Estimated from Various Sources

Level 1	Level 2	Level 3	Type	Area (km²)	Remarks
	32 Shrubs and/or herbaceous associations	321 Natural grasslands	Meadows	0.676	
		322 Moors and heathland	Commons	6.520	
		324 Transitional woodland/shrub	Moors	0.787	
			Heather	2.067	
				4.314	
	33 Open spaces with little or no vegetation	331 Beaches, dunes, and sand plains	Meadows	1.617	Limited importance
4 Wetlands	41 Inland wetlands	411 Inland marshes		—	
		412 Peat bogs		—	
	42 Coastal wetlands	421 Salt marshes		—	
		423 Intertidal flats		—	
5 Water bodies	51 Inland waters	512 Water bodies		0.532	
	52 Marine waters	521 Coastal lagoons		—	
		523 Sea and ocean		—	
				111,656	Total

Note: Roads I and Roads II are paved/unpaved roads, length estimated from MapInfo data from municipality—333.349 m and 147.410 m. Data for various types of larger nature according to CLC are identified from municipal data.

3.3 Quantifying the Infrastructure

Human societies possess a series of structures, including buildings (houses), roads and harbours, that are necessary for sustaining a variety of societal activities. An attempt of collecting data in this area is demonstrated in the following.

3.3.1 The Urban Fabric vs. the Rural Zone

The area of Samsø is divided into 6,754 entries in the land registry. The entries in the land registry account for a total area of 112,23 km², which is slightly less (98.2%) than the area of the island as a whole (114.26 km²).

The areas of the various zones as recorded in the registries and subdivided into the land registry zone types are shown in Table 3.3.

As seen in the table, the area of the island is dominated by the rural landscape which takes up 95.65% of the area, leaving 4.33% to urban, rural/urban, and summer cottage areas (0.02% of the total area has not been identified in the land register).

3.3.2 Dividing the Infrastructure

In the public registry of buildings and housing units in the country (in Danish *Bygnings- og Boligregistret*, aka BBR) a total of 8,787 buildings can be identified, comprising a total area of 854.722 m². Since the total number of entries is 6,754 (see the earlier discussion), it follows that the average land registry entry has 1.3 buildings on it. For unknown reasons, the area of 51 buildings—corresponding to approximately 0.6% of the total—was not stated.

The buildings can generally be assigned to four sectors:

TABLE 3.3

The Number of Various Zones Given in the Land Register Together with the Respective Areas and Relative Contribution to the Landscape on the Island of Samsø

Zone Text	Number	Area (km²)	Pct.
Rural zone	4,665	107.34	95.65
Summer cottage area	792	1.56	1.39
Rural and summer cottage area zone	35	0.75	0.67
Urban zone	678	1.63	1.45
Urban and rural zone	473	0.93	0.83
Unknown zone code	111	0.02	0.02
Totals and percentage	6,754	112.23	100

1. A *public-sector* infrastructural component which includes buildings that serve as housing facilities to the municipal administration and services, as well as public buildings such as cinemas and sports facilities (not necessarily owned by the municipal authorities)

2. A private-sector component consisting of the housing facilities as well as additional buildings belonging to private domestic activities, such as garages and huts. The housing facilities of farmers are included in this group (not considered as belonging to business activities).

3. The remaining buildings that belong to farming activities: storage buildings for livestock, crops, farm machinery and so on, that is the structures needed to run the farming activity

4. Buildings belonging to industrial and trade activities, ranging from actual production facilities to storage houses, offices and the like

The overall composition in fractions is shown in Figure 3.1. Data have been provided by the municipality.

The figure indicates the predominant role of private housing units, corresponding to almost 55% of the total built area, followed by agricultural buildings (29%, excluding housing units); industry, trade and commerce (almost 11%); and the public sector (5%).

3.3.3 The Public Sector

The public sector infrastructure is here defined as

- buildings and other structures such as roads, that are either used or maintained by public administration, in this case the municipality of Samsø, as well as

- structures that are necessary for supplying society with the necessary services and commodities, (mainly consisting of storage facilities for buses, road-mending machinery and other equipment) and

- buildings and structures used by the public for social purposes like cinemas, libraries and meeting places.

The number of buildings and the cumulated area represented by buildings in each group is shown in Table 3.4.

The public-sector buildings may be divided into five sub-groups:

1. Buildings belonging to production facilities
2. Buildings related to public services of the society
3. Buildings for educational purposes (schools, kindergartens)
4. Buildings containing recreational facilities
5. Buildings for other functions and institutions not belonging to the preceding sub-groups

A diagram showing the distribution among the five sector functions is shown in Figure 3.2.

It should be noted that part of the public administrative facilities has been registered under the "industry, trade and commerce" sector buildings.

3.3.4 Private Sector

This sector comprises the infrastructure necessary to house private families and contain their households as well as adjacent and necessary structures housing other consumables and use commodities and utensils such as gardening equipment and tools., bicycles and cars.

As already mentioned, the housing facilities of farms are included in this group as well as housing facilities only used for leisure purposes and holidays.

The private-sector building area can be divided into a number of subgroups. First, there are the buildings used for housing proper, that is providing the living space of families and individuals. In addition to this, on the grounds of many private houses there will be additional buildings used for a variety of (non-residential) purposes, such as sheds and garages. A final subgroup of private-sector buildings is represented by what we will call second homes: buildings used mainly at weekends and holidays.

TABLE 3.4

Distribution of Public Infrastructure among the Types Found in the Public Register over Buildings

House Type/Sector Public/Societal		Number of Buildings (#)	Cumulated Area (m²)
4	Building for cinema til biograf, theatre, occupational exhibitions, library, museum, church and alike	37	7,923
4	Other building for recreational purposes	72	3,349
1	Electricity-, gas-, water- or heating or combustion plant, etc.	45	8,457
3	Building for teaching and research	41	12,200
2	Care centres (nursing homes, old people's home children's homes, youth centres)	9	4,595
5	Building for other institutions, herunder barracks, prisons and alike	3	1,059
5	Residence halls	12	3,421
2	Building day care centres	9	1,682
4	Building in connection to athletic sports (club houses, sport centres, swimming baths and alike	16	6,694
2	Building for hospital, homes, maternity homes and alike	5	912
	Subtotals	249	50,292

The distribution of these buildings among the three sub-groups is shown in Figure 3.3.

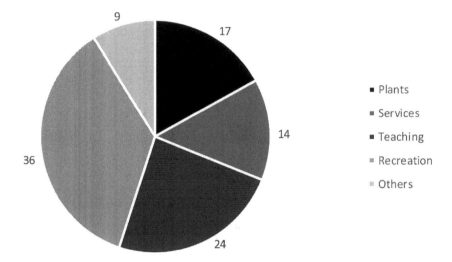

FIGURE 3.2

Diagram showing distribution (pct.) of the buildings considered to belong to the public sector (w/o administration) among five major functional groupings: production plants, services, education, recreation and other purposes.

TABLE 3.5

Distribution of Private (Residential) Infrastructure among the Types Found in the Public Register over Buildings (BBR)

Sector: Private Households	Type of Building	Number of Buildings (#)	Cumulated Area (m^2)
Housing	Multi-storey buildings (multi-family house, including two-family houses (horizontal separation between units)	82	14,814
	Non-detached or double house (vertical separation between units)	181	29,040
	Detached single-family house	1,843	201,388
	Other building for year-round occupancy	119	7,872
	Farmhouse on agricultural holding	358	52,028
Cabins/garages	Outhouses	2,199	90,608
	Garage with space for one or two vehicles	647	23,731
	Carport	452	12,844
Cottages	Summer houses	993	68,434
	Building for holiday purpose, etc., other than summer houses (holiday camps, hostels and alike)	32	4,911
	Subtotals	6,906	505,670

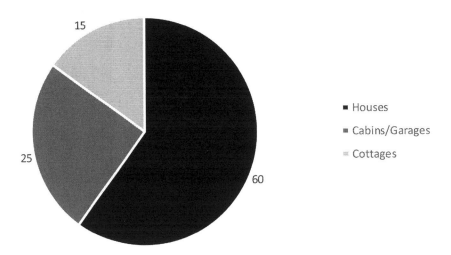

FIGURE 3.3
The distribution of buildings within the private sector among three major groups: houses used for everyday living, adjacent buildings like outhouses, sheds, cabins and garages, and secondary houses used during summertime and holidays

3.3.5 Agricultural Sector

The infrastructure here is concerned with agricultural buildings other than farmhouses (included in the private households); that is this sector covers the additional buildings necessary for the agricultural activities on a farm, that is structures concerned with the storage of crops after harvest, the storage of agricultural machinery and the raising of animals.

The amount of infrastructure considered to belong to the agricultural sector is shown in Table 3.6.

As seen from Table 3.6, this sector, in fact, comprises includes quite a heterogeneous set of buildings, used both for agriculture and industry, as well as for other related activities.

Meanwhile, most of the commercial activities undertaken on the island are closely linked to agriculture, and in this case, the errors and uncertainties introduced by attempting a more accurate partitioning of this type of infrastructure are considered to be of minor importance.

3.3.6 The Industrial Sector

The infrastructure in this sector is used for various kinds of activities undertaken within our societies, ranging from industrial activities proper, via trade and commercial activities to structures providing a wide range of services that are vital to the existence and orderly functioning of a society.

As the last two groups listed in the register are both related to transport activities, they may represent a subdivision of the buildings in this sector

TABLE 3.6

Distribution of Private (Residential) Infrastructure among the Types Found in the Public Register over Buildings (BBR)

Sector: Agriculture	Building Types	Number of Buildings (#)	Cumulated Area (m²)
	Buildings for commercial production related to agriculture, horticulture, extraction of raw materials and alike	1,112	260,365
	Other building for agriculture, industry, etc.	57	10,464
	Subtotals	1,169	270,829

TABLE 3.7

Distribution of Buildings within the Industrial, Trade and Commerce Sector among the Types Found in the Public Register over Buildings (BBR)

Sector: Industry and Trade		Number of Buildings (#)	Cumulated Area (m²)
	Building for offices, trade, stocks, including public administration	174	39,726
	Building for hotel, restaurant, laundry, hairdressers and other service activities	79	11,422
	Building for commercial production pertinent to industry, trade, etc.	106	40,628
	Transport- og garage plantsanlæg (carrier activities, airport building, railway buildings, car park buildings).	15	4,396
	Other building for transport, trade, etc.	38	3,978
	Subtotals	412	100,150
Not included	Un-identified	51	0

into four groups belonging to types of activities: offices, service functions, productional and transports sector facilities.

The distribution of the areas among the groups is found in Figure 3.4.

The areas of the industrial sector are dominated by two major sectors, offices and production, which accounts for 81% of the area, and services and transport, which accounts for the rest (19%).

3.3.7 Summary of the Distribution by Area of the Sectors

The space taken up by the various sectors and activities varies considerably over time. The data presented here are taken from the municipal records, and this seems to be the optimal way of gaining insight into the overall activities and the space used for these activities for any region.

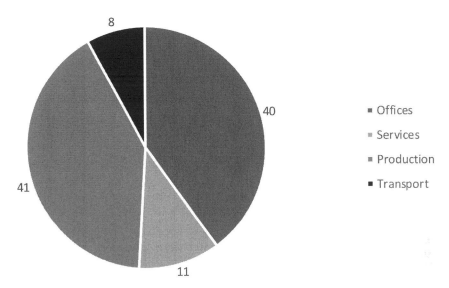

FIGURE 3.4
The distribution of areas of buildings within the sector of industry, trade and commerce and their respective functions within the sector

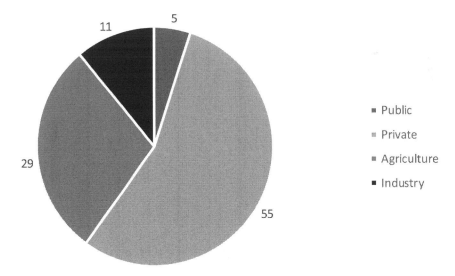

FIGURE 3.5
The overall distribution of built areas among the sectors

The areas occupied by the various sectors are recorded in Tables 3.4 through 3.7 and are also shown in Figure 3.5.

The private sector is seen to be far the largest sector in terms of building space (55%), followed by the agricultural sector. These two sectors together take up 84% of the built area on the island. The industrial and related activities and the public sector account for the remaining 16% of the area.

3.4 Work Energy of Energy and Materials

As indicated in Chapter 2, significant amounts of work energy are included in different forms in the numerous activities undertaken by a society. They may be considered separately, however, and for many municipalities, the first step taken to gain knowledge about the relative importance of the various activities is to establish an overall energy budget. To acquire similar, additional knowledge about the distribution and use of materials or substances in our society is, in most cases, very difficult and requires information from additional sources. This aspect needs to be improved in the future.

In this section, we consider work energy in three forms: the work energy content of *energy flows*—covering everything from the work energy in solar radiation to the chemically bound work energy in fuels, as well as the work energy potentials of thermal inputs:

- The work energy of more or less pure chemicals and elements—analysis of which is often referred to as substance flow analysis
- The work energy of *materials* in society; many of the materials are conglomerates of various chemical compounds and elements, and the final goods produced may contain a number of components. Their production and composition also require the expenditure of work energy.

The first type of work energy has been well documented in the scientific literature, particularly after the establishment of the *International Journal of Exergy* and the *Journal of Cleaner Production*, the *Journal of Industrial Ecology* and other related engineering journals. For an overview of the concept, reference should be made to Sciubba and Wall (2007).

The two latter types of work energies have been documented by a mixture of LCA and MFA studies, also published in the previously mentioned journals and demonstrated for instance in Baccini and Brunner (2012).

For chemical compounds and elements, in particular, useful values may often be found in the works of Szargut and coworkers (Szargut et al., 1988; Szargut, 2005), as well as in other sources. For many complex organic compounds, the work energy content can be calculated using the procedures described by this author or using the "exergy calculator" introduced since 2018 at the Exergoecology site (www.exergoecology.com/excalc).

TABLE 3.8

Exergy Values of Some Elements and Materials in Our Society

Material	Value	Reference	Remark
	(kJ g^{-1})		
Metals			
Iron	6.7	Ignatenko et al., 2007	Chemical exergy
Copper	147.4	Koroneos and Kalemakis, 2012	CEC value
Copper	2.1	Ignatenko et al., 2007	Chemical exergy
Zinc	198.9	Koroneos and Kalemakis, 2012	CEC value
Zinc	5.2	Ignatenko et al., 2007	Chemical exergy
Aluminium	250.2	Koroneos and Kalemakis, 2012	CEC value
Aluminium	27	Ignatenko et al., 2007	Chemical exergy
Building Materials			
Concrete	1.7	Koroneos and Kalemakis, 2012	CEC value
Wood			
Steel	47	Koroneos and Kalemakis, 2012	CEC value, reinforced
Bricks	2.7	Koroneos and Kalemakis, 2012	CEC value
Plaster/plasterboard	7–9	Koroneos and Kalemakis, 2012	CEC value
Ceramic tiles	3.2	Koroneos and Kalemakis, 2012	CEC value
Calcium carbonate	16.3	Gaudreau, 2009	
Other			
Paper/cardboard	59.9	Koroneos and Kalemakis, 2012	CEC value
Glass	21.1	Koroneos and Kalemakis, 2012	CEC value
PVC	82	Koroneos and Kalemakis, 2012	CEC value
Rubber	45.5	Ignatenko et al., 2007	Chemical exergy

Note: CEC = Cumulated exergy value.

3.4.1 Work Energy of Energy and Matter

In order to determine the work energies of stocks and flows of material/substances in our society, it is necessary to know the work energy of elements and the elemental compositions of the product or sub-products in question or of the chemical compounds.

Work energies for a variety of activities and processes, including the production of energy and of material goods, may be found in the works of, among others, Dincer and Rosen (2007), from which reference some conversion factors have also been taken. This scientific area is continuously expanding as our attention is drawn to the sometimes unexpectedly high environmental burdens created by activities that we take for granted in our society. This is the case for construction materials such as cement, for example, the production of which is both very energy-intensive and threatens to use up our known deposits of sand and gravel, resources previously considered to be unlimited.

The work energies of composite materials may also be found in the existing literature, but as they are conglomerates of real elements and other compounds they are often represented by cumulative values (see also Table 3.9) which represents the total embedded exergy costs in time (all production phases) and space (composing elements).

Initially we will be concerned mostly with the energies used to drive our society which hitherto have been derived mainly from non-renewable resources, such as fossil fuels, as well as from the renewable sources that have recently been developed, such as wind turbines and photovoltaics.

The work energy content of energy from the previously mentioned sources may be calculated by multiplying the energy content by a factor expressing the work energy density (relative exergy content) of that specific type of energy, that is the work energy of energy type i (WE_i) may be described as

$$WE_i = \xi_i \cdot E_i, \tag{3.1}$$

where ξ_i is the fractional content of work energy in energy type i and E_i is the amount of energy of type i.

In the case of radiative energy, nuclear energy and electricity, as well as very energy-dense chemical compounds (where the energy content is high and compact) the value of ξ_i is 1 (or close to 1) meaning that these are energies with an inherently high work-energy potential. This, in fact, is part of the reason why fossil fuels and their derivatives are interesting.

For other chemical energies, the values of ξ_i are lower, as also indicated in the previous chapter. Ultimately, all energy is dissipated and all work energy potential exhausted, and we end up with heat at ambient temperature with a value of ξ_i (and thus also a work energy potential) of 0 (zero).

3.4.2 Work Energy of Construction Materials and Houses

In order to take the final step of calculating the total work energy bound up in for example buildings with a reasonable degree of certainty it is necessary to know two additional things. First, we need to know the relative composition of buildings in terms of fractions of various types of building materials. Second, we need an estimate of the work energy content of the building materials, which are often composed of many constituent materials.

Having determined this, we will be able to make a fairly reasonable estimate of the work energy content incorporated in a given building infrastructure, calculated as

$$WE_{bt} = \sum_{bt} M_{mat,bt} \cdot WE_{mat}, \tag{3.2}$$

where WE_{bt} is the work energy contained in the infrastructure of a specific type of building (bt), determined as the masses of materials (i components)

TABLE 3.9

Relative Composition of 6 Types of Houses Assessed for 11 Fractions of Materials (all values in kg m^{-2})

Materials of Type	Stones and Sand	Concrete and Mortar	Tile and clinker	Metals	Wood	Cardboard and Linoleum	Mineral Wool and Fiber	Plastics	Glass	Bitumen Products	Paint etc.
Detached house	71	672	192	7	44	0	14	2	2	2	2
Non-detached	103	747	168	7	40	0	13	2	2	2	3
Multi-story	69	855	141	17	36	0	9	2	3	1	2
Farm buildings	276	722	62	17	23	0	4	1	0	1	0
Industrial buildings	158	934	56	20	13	2	8	1	1	5	1
Office and admin.	156	918	62	28	18	2	7	4	2	2	2

Source: Kaysen and Petersen (2010).

TABLE 3.10

The Exergy Contents of Various Materials Commonly used in Building and Construction (values in MJ kg⁻¹).

Fraction	Work Energy Value (MJ kg⁻¹)	Comment	Reference
Stones and sand	1.7		
Concrete and mortar	1.7		
Tile and clinker	0.75		
Metals	200	No distinction possible	
Wood	18.7	As detritus	
Cardboard and linoleum	59.9	Quite different materials in origin	
Mineral wool and fiber	21.1	As glass	
Plastics	91.9		
Glass	21.1		
Bitumen products	40	Oil products approx.	
Paint, etc.	1.7		

contained in that building type, and WE_{mat}, which is the work energy density of the specific material.

Thus, to assess the work energy contents of buildings, it is necessary to know the areas of the different building types, their average composition in terms of building materials and the respective work energy densities of these materials.

An assessment of the composition of six types of buildings typically found in Denmark can be found in Kaysen and Petersen (2010). Together with other sources of data, it is possible to give an estimate of the work energy in the sectors and implicitly what is needed to run them in terms of remodelling and so on (Nielsen, 1993).

The types given are

1. detached houses,
2. non-detached houses,
3. multi-storey buildings,
4. farm buildings,
5. industrial buildings, and
6. offices and administrative buildings.

The compositions have been assessed in terms of 11 fractions.

Table 3.11 shows the relative compositions of 6 types of houses assessed in terms of 11 fractions of materials: all values in kg m⁻² (Kaysen and Petersen, 2010).

TABLE 3.11

Table Showing the Areal Exergy of 6 Types of Buildings Distributed among 11 Fractions of Materials Found by Combining Tables 3.9 and 3.10 (all values in MJ m⁻²).

Materials of Type	Stones and Sand	Concrete and Mortar	Tile and Clinker	Metals	Wood	Cardboard and Linoleum	Mineral Wool and Fiber	Plastics	Glass	Bitumen Products	Paint etc.
Detached house	120.7	1,142.4	144	1400	822.8	0	259.4	183.8	42.2	80	3.4
Non-detached	175.1	1,269.9	126	1400	748	0	274.3	183.8	42.2	80	5.1
Multi-story	117.3	1,453.5	105.75	3400	673.2	0	189.9	183.8	63.3	40	3.4
Farm buildings	469.2	1,227.4	46.5	3400	430.1	0	84.4	91.9	0	40	0
Industrial buildings	268.6	1,587.8	42	4000	243.1	119.8	168.8	91.9	21.1	200	1.7
Office and admin.	265.2	1,560.6	46.5	5600	336.6	119.8	147.7	367.6	42.2	80	3.4

The relative exergy content (mostly cumulative) of each fraction (MJ kg^{-1}) is taken from the literature and is shown in Table. 3.10.

The work energy (exergy) content as density, that is as MJ m^{-2}, for the various building types given earlier can now be calculated (information partly derived from Kaysen and Petersen [2010] and Nielsen [1993]).

3.4.3 Work Energy of Chemicals and Elements

The work energy of elements may be found in the works of Szargut (Szargut et al., 1988; Szargut, 2005). For other compounds synthesized from these elements, a strict methodology for calculating the work energy content is described in the same references.

Some of the methodological steps taken here are quite demanding in the sense that they require a high degree of chemical insight, for instance, with regard to the synthetic pathway(s) of the molecular components of a given chemical, together with a set of thermochemical information for all—or most—of the compounds involved in the synthesis.

It is therefore quite a time-consuming procedure—not to say a tedious task—to carry out such calculations for all the chemical compounds found in our societies today. More than 100,000 compounds are registered as chemicals for use within the European Union. This set must be reduced to cover the compounds most commonly used, that is used in the highest quantities.

For this type of analysis, it may also be feasible to reduce the need for data sets for instance by considering and involving only the chemicals actually used within a specific society. For example, Danish agriculture involves a subset of possible additives corresponding to about 80 to 100 compounds used as fertilizers, herbicides and pesticides.

The expansion of the project to other countries would lead to an increasing demand for knowledge about compounds, which would involve the development of a database proper.

In this context, therefore a compromise has been chosen, namely using the average values for groups of compounds—and in some cases even relying on cumulated values.

3.5 Other Elements of Methodology

At the beginning, we already mentioned the need to somewhat deconstruct the society under consideration into smaller and more controllable units. This is because an aggregated overview of energy or work energy balances may well serve as a good indicator of overall performance but leaves us with little or no chance of identifying the underlying reasons for a given situation and revealing where action can or must be taken.

In order to determine what happens in a society and to indicate more exactly the degrees of efficiency and the related potential for the optimization of various functions, it is convenient to take a more hierarchical view of our society.

As opposed to society we view nature as a separate component which is partly dependent on the activities of society (e.g. environmental management) and potentially impacted by the same activities (e.g. exploitation, pollution).

In the present case, therefore, the study area (Samsø) has been divided into 7 (if waste is included as a separate sector) sectors which are illustrated in Figure 3.6.

In this outline, we have identified the following sectors as being commonly present and important for establishing the work energy balance of any society:

- The energy-producing sector (ENER)
- The public sector (PUBL)
- The private sector (PRIV)
- The agricultural sector (AGRI)—which may be split into four activities: crop production, livestock production, forestry and fishery
- The industrial sector (INDU)—also including commerce and trading
- Nature (NATU)—all ecosystems

The preceding abbreviations are used in the flow descriptions later in this chapter. The method has, in fact, been developed in this manner in order to facilitate its transfer to a software based on a web page system allowing municipalities to enter numbers and carry out sustainability calculations of the costs of activities for their own society (e.g. via spreadsheet).

In the following we briefly discuss some methodological issues related to this deconstruction of society, in particular issues related to the requirements for and availability of data.

3.5.1 Geographical Issues and Landscape Perspective

As mentioned earlier, the area data for the analysis have been derived from a variety of data sources that are not always linked in an optimal way; sometimes the linkage does not exist at all. For example, the relationship among land ownership, land use and related activities needs to be established much more precisely in order to refine the method.

Despite this, some major steps have been taken which make it possible to produce a work energy analysis of a society that can be used to identify major issues. It can also be used as an overall guide for the initiation of further studies and applications.

a society

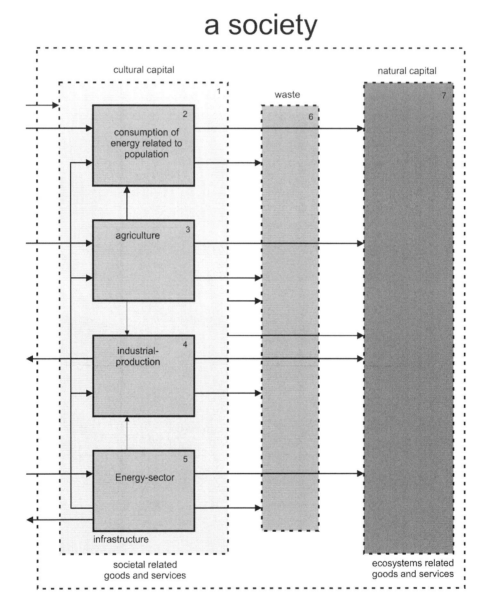

Energy-balances and -fluxes

FIGURE 3.6
Diagram showing the partitioning of activities on Samsø—or any society—combining the respective sectors

In the meantime, most municipalities today subscribe to systems that combine fundamental sets of spatial data and as mentioned, in Europe these systems, to a great extent, relate to the CLC system. Within municipalities, the detail is often much higher than the requirements set up by the CLC system. The available data include information on the variety of structures mentioned in Sections 3.1 through 3.4, typically buildings and the actual usage of the areas, as well as roads and other service structures belonging to society.

Agriculture and its activities make up a whole sub-system of its own. The regulatory function—at least for crops—lies at present with the Ministry of Food, Agriculture and Fisheries of Denmark. In some cases, it might be easier to rely on information from the supervising or controlling authorities, but this may vary as it depends on the size of the structure under analysis. In the case of Denmark, it depends on whether the analysis deals with a municipality or a region; resolution and chances of identifying details get lost.

Data for the industry, trade and commerce sector will pose problems for a determination of this kind. First of all, the industrial sector is normally highly diverse—much more diverse than in the case presented here. Furthermore, the relative role of the many types of activity may vary a lot even within a rather confined area. With the industrial activities the elements, material substances and possible pollutants will also vary, and therefore, the work energies involved will vary even more. The (work) energy intensity and the supply demands of the processes will also vary. All this contributes to the (im)possibility of making a systematic approach to this sector.

The more feasible way is likely to be one that takes its starting point in industries or factories that already report their consumptions to the authorities through some sort of green accounting system. As the legislative requirements for factories and companies are often based on size as one parameter, it should be possible to get a quantitative picture. Such a study should be accompanied by the setting-up of a database covering the most normal processes and compounds involved in industries that are of interest.

Such a starting point combined with an extensive database of work energy values and correlations to typical values from LCA/MFA studies is seen as being necessary but probably also sufficient to provide an overview of the relative importance of activities in any of the sectors under consideration.

Evaluation of the importance of nature and its inherent ecosystem services is also possible within the framework, but the authorities responsible for estimates of the geographical extent of such systems may change with time, and this in turn may well influence the quality (mainly resolution) and the availability of data. Technology nowadays offers the possibility to monitor ecosystems quite precisely in terms of biomass and production by use of for instance remote sensing as a "quick and dirty" approach, but such techniques are at the moment hardly accessible to all authorities. Anyways, in this case of giving an approximate estimation of extension and types of ecosystems together with their respective biomass, most municipalities should be able to provide data at an adequate degree of resolution to carry out an analysis as presented here.

What is even more crucial is the availability of data for the various types of ecosystems so as to make it possible to couple their spatial extents to information about the biomasses in the various pools in the systems and their values in terms of work energy or even eco-exergy. Much of the present overall knowledge is rather old and imprecise in its estimates. Furthermore, the data are derived from generalized types of ecosystems, possibly from other geographic and climatic conditions and not always easily related to the local ecosystems under consideration. Eventually, specific local knowledge of biomasses (of various trophic components) and processes such as net primary production and respiration would represent the ideal situation, but it will probably rarely be the case that all such information will be available.

Attempts to get more adequate information, such as more accurate descriptions of the ecosystems and data that are relevant for the systems today, may represent a severe limitation, even when only attempting to obtain a relatively precise estimate of activities and pools of work energy. As conversions to eco-exergy involve conversion factors dependent on the level of organisms involved this increases the need for data about the level of biodiversity. The conversion factor in itself only has the effect of magnifying the differences found. In some cases, these problems can be solved by literature studies, probably involving so-called grey literature, but in others, there may be a need for actual studies of the specific organisms present in the system.

Some pools that may be interesting to analyze further in the future are the pools of organic matter and carbon associated with the various landscape types which have not been investigated well enough to make a more accurate estimate than the one made here. This is important knowledge as information about actual carbon sequestration and possible alternatives is important when evaluating sustainability in a climatic context. The same applies to the zoological component of an ecosystem, animals in the widest sense, including both invertebrates and vertebrates. In the case of Samsø, mammals, and deer, in particular, could be of interest for further study as their present density is believed to be close to the carrying capacity of the island. Part of the animal component together with bacteria and fungi may play an important role in the storage or recycling of elements in a given ecosystem.

3.6 Work Energy Methods in the Sectors

In this section, we give an outline of how work energy budgets can be developed for the sectors mentioned earlier. The basic outline involves an inventory of the stocks in the system under consideration and their associated incoming and outgoing flows, all considered and accounted for in terms of work energy

An additional sector is introduced which is referred to as the energy sector (Section 3.6.1). This is found to be convenient because an energy budget is often already reported in many municipalities. Such inventories usually serve to give an overview of the energies imported to the municipality divided into their various types, their usage and distribution and the various types of energy consumption in the society.

In cases where this type of inventory has already been produced, it makes a good starting point for an initial identification of the important sectors and processes, as the only action initially required is a conversion of the energy form to work energy. In the following, we review the sub-sections introduced in Section 3.5 and discuss what will normally need to be considered within each section. A schematic representation of the society was presented in Figures 3.1 and 3.7.

3.6.1 Energy Production and Consumption—Methodology

In general, this section deals with the energy balances of a municipality, county, nation, regions or, for that matter, any sort of society. On one hand, this fundamental part of an analysis starts off in the same way that this type of analysis is normally undertaken, that is an energy accounting exercise which in principle corresponds to a first-order analysis. On the other hand, the approach taken here adds to this an attempted conversion of the traditional energy budget scheme to also include second-order considerations by converting the energy balance to a work energy (exergy) balance. This may sound trivial but is not always as easy as it appears. For the sustainability evaluation, we have to start grouping energies into fractions according, for instance, to their dependency on renewability of resources or their further availability to our system after any conversion process.

A typical starting point for researchers embarking on this part of the analysis is to identify the energy budget. Such an account is often produced already by local authorities in accordance with existing legislation.

The flows of the energy sector vary considerably in character. This not only is a direct consequence of the variety of inputs necessary to run a society but also arises from the many transformations of energy quality that result from the use of energy in general. Relevant aspects here may include information about the sources of energy proper, dependency on fossil fuels, the chances of making efficient links between different energy systems and so on.

The necessary energy inputs may be obtained in quite diverse ways. Traditionally, unless we are dealing with countries or regions which are self-sufficient in fossil-fuel resources, all areas will need either to import the amounts of energy needed to run their society or to set up production facilities for alternative forms of energy. Here we will avoid the debate about resources vs. reserves, since, at this point, we have no intention of including economic considerations.

Usually, the energy running most industrialized societies today is based either on fossil fuels, i.e. coal, oil and gas, or on nuclear power plants where atomic energy is transformed into electricity. The energy efficiency of the various raw materials used is quite similar, whereas work energy efficiencies may be quite different. This fact underlines the need for a new approach, as pointed out by Sciubba and Wall (2007). It is important to note that every time a conversion between energy forms occurs, work energy is lost, even though the work energy content of the product is actually raised.

Many other derivatives of fossil fuels, mainly from crude oil, are needed for other purposes than electricity and heat production. Most of these are direct results of cracking of the crude oil into gasoline, diesel, kerosene and so on. Other material products such as bitumen derivatives and plastics find their origins in fossil fuels and are likewise dependent on this supply.

Recently, with the recognition of the "peak oil problem" (Hubbert, 1962; Campbell, 1997), a search has begun for new ways of safeguarding the energy and material basis of our society. In sustainability terms, this means that we have to invest research and educational efforts into preserving sufficient resources of fossil fuels and a variety of material compounds until we have developed technologies efficient enough to support the existence of future generations. This assumes, of course, that we take the messages from the definition of sustainability and the peak production discussion—for all resources—seriously.

Among the new technologies proposed, we will find the following:

1. Attempts to extend the time during which we can continue to use existing technologies
2. Attempts to convert the platform of existing techniques to be more sustainable using relatively low-tech technologies (e.g. ecological engineering, eco-technology)
3. Attempts to convert the existing energy supply platform to be based on more sustainable high-tech forms, (hydrogen power, fusion plants), most of which unfortunately are at quite an immature stage today, although the proposed technologies are considered to have great potential

Our attempts to expand existing technology range from increasing the efficiency with which we can exploit and deplete the existing fossil-fuel-based resources (oil, coal and gas) to the development of new but simple technologies that allow us to continue a downstream use that will still be based on fossil fuels. This has the obvious advantage that many conversions are avoided and helps to exploit the last drop of still available energy, following the previously mentioned MEP principle (Jaynes, 1957a, 1957b).

In the end, such approaches may in fact be dangerous, as it will be tempting to continue a path following "business as usual" rather than to invest the necessary research in the development of methods for the future. This means that we are basically ignoring the fact that the resources behind this approach are also finite, for instance when we choose to expand our fossil-fuel-based technology with the exploitation of similar resources such as tar sand and shale gas. At the same time, we must recognize that the present technologies for exploiting resources like tar sand not only are inefficient but also pose a threat to the environment in general by damaging forested areas or discharging toxic substances to ground-water.

In terms of shaping a transition towards a sustainable state, much can be done by making our society run on less sophisticated but well-tested technologies such as wind power and hydropower, which represent technologies that have already been known for millennia. This is almost like re-inventing the wheel and going back to fundamentals. But intermediate stages like photo-thermal devices and heat exchangers may also turn out to be beneficial for working towards higher sustainability, to say nothing of simple exploitation of biomass by burning. Here the discussions of the problems concerning the carbon balance of this technique are ignored, but clearly they need to be settled and the pros and cons to be weighed against each other.

Conversion of society by means of high-tech-based solutions is possible but inherits some basic disadvantages like, for instance, solving the fossil-fuel problems of the transport sector by the introduction of electric vehicles. Solutions are often based on finite resources, they often require a high consumption of energy for their construction which at present is usually derived from finite material resources, their wastes are often problematic and so on.

One argument taken from Ayres (1999), amongst others, would be that if we eventually in the future can convert our energy supplies to be based on infinite resources such as solar radiation, we can then ignore the problem of energy supply, meaning that we ought to invest much more in photovoltaics. The argument is that solar radiation reaches Earth in amounts that—if captured and exploited efficiently—are far greater than our current needs. We only need to develop the technology to exploit this situation fully. In this way, we get sufficient energy not only to replace fossil fuels but also enough to close material cycles in a "cradle to cradle" manner as implicitly demanded by the circular economy.

3.6.1.1 Energy Production and Consumption

According to the basic layout given in Chapter 2, the energy sector will be divided into energy- and matter-bound energies and further divided into sustainable vs. non-sustainable components. The terms employed to symbolize stocks, in- and outflows, energy vs. matter, renewable vs. non-renewable within the energy sector are suggested to be as follows—this is in order to facilitate potential transfer to a database-like system:

For Stocks:

Renewable Energy-Bound Exergy Stocks—REBES_ENER

Non-Renewable Energy-Bound Exergy Stocks—NEBES_ENER

Renewable Matter-Bound Exergy Stocks—RMBES_ENER

Non-Renewable Matter-Bound Exergy Stocks—NMBES_ENER

For Inflows:

Renewable Energy-Bound Exergy Inputs—REBEI_ENER

Non-Renewable Energy-Bound Exergy Inputs—NEBEI_ENER

Renewable Matter-Bound Exergy Inputs—RMBEI_ENER

Non-Renewable Matter-Bound Exergy Inputs—NMBEI_ENER

For Outflows:

Renewable Energy-Bound Exergy Outputs—REBEO_ENER

Non-Renewable Energy-Bound Exergy Outputs—NEBEO_ENER

Renewable Matter-Bound Exergy Outputs—RMBEO_ENER

Non-Renewable Matter-Bound Exergy Outputs—NMBEO_ENER

From the investigations carried out in connection with such a project—the mapping of energy and matter flows converted into work energy (exergy)—it is possible to elucidate and evaluate the degree of sustainability of the sector.

3.6.1.2 Energy Stocks in the Energy Sector

a. Renewable Energy-Bound Exergy Stocks (REBES_ENER)

In fact, no stock facilities have really been developed to store the work energy from renewable energies. This is in spite of the fact that much infrastructure is often dedicated to the partitioning/ conversion of energies prior to distribution and to ensure the right quantities and forms (see Chapter 1). Several recent candidates exist for storing energy, for example various types of batteries including hydrogen batteries, but so far, the best option seems to be to store it in reservoirs for hydropower or—on a shorter time scale—to store it in batteries in electric vehicles.

b. Non-Renewable Energy-Bound Exergy Stocks (NEBES_ENER)

Stocks or stores of work energy in non-renewable resources arise every time organizations stock fossil fuels on a larger scale, for example at gasoline stations or in larger storage facilities in harbours for ships. The issues are in most cases probably negligible. In fact, this type of storage is mainly relevant to oil-, coal- or gas-producing countries where large stocks may be found.

c. Renewable Matter-Bound Exergy Stocks (RMBES_ENER)

These stocks consist mainly of biomass, such as bales of straw, stored timber, tree trunks, woodchips and pellets, among others. The stocks can also consist of products produced from renewable matter: biogas, bioethanol, and so forth.

d. Non-Renewable Matter-Bound Exergy Stocks (NMBES_ENER)

To this group belongs all stocks of finite materials (elements found in ores— other than coal- and oil-related stocks). Items such as metals are placed in this group and products derived from non-renewable energy stocks such as different types of plastics.

3.6.1.3 Energy Inflows to the Energy Sector

e. Renewable Energy-Bound Exergy Inputs (REBEI_ENER)

A list exists of the possible inputs and production of renewable (and there-fore sustainable) inputs of work energies. To this group belong the inputs from wind power, hydropower (reservoirs and tidal power plants) and wave power (if this technology proves viable). Input from solar radiation, such as photovoltaics and solar heating panels, should also be included in this group.

f. Non-Renewable Energy-Bound Exergy Inputs (NEBEI_ENER)

The inputs normally originate in the form of oil, coal or gas, but recently also the exploitation of tar sand or shale oil/gas has been added, as discussed in the introduction.

g. Renewable Matter-Bound Exergy Inputs (RMBEI_ENER)

This input predominantly must include the inputs of energy bound in bio-mass. In many cases, in order to be considered sustainable, these inputs must be restricted to include only an amount corresponding to the regional annual growth in biomass occurring as a consequence of photosynthesis. To estab-lish the sustainability of this input, it is also necessary to consider where and how this resource has been produced: the more local in origin, the better.

h. Non-Renewable Matter-Bound Exergy Inputs (NMBEI_ENER)

This category again involves the necessary inputs of finite elements such as metals to a given system. Again, it is necessary to address the origin of the materials. Reuse and recycling are typical ways of reducing the need for these inputs.

3.6.1.4 Work Energy Outflows from the Energy Sector

i. Renewable Energy-Bound Exergy Outputs (REBEO_ENER)

A system that produces more than its own consumption of energy by means of sustainable/renewable technologies may regard its exports as renewable outputs of work energy, for example the surplus electricity produced by wind power on Samsø or the excess electricity produced by hydropower in Norway and Sweden.

j. Non-Renewable Energy Bound Exergy Outputs (NEBEO_ENER)

Regions or countries that for geological reasons have large deposits of fossil fuels (e.g. the Organization of the Petroleum Exporting Countries [OPEC]) may become exporters of this type of work energy.

However, all energy consumption leads to an output in terms of a loss of work energy. According to the second law of thermodynamics, all activities must have a cost, namely the inevitable formation of heat—which represents the major loss in this group. Several techniques exist and are widely deployed today to minimize these losses, which will need to be compensated by increased energy input.

k. Renewable Matter-Bound Exergy Outputs (RMBEO_ENER)

Renewable matter-bound work energies occur whenever a system has exports of goods originating in biological materials or production. Although renewable, such an export is not necessarily sustainable, since excessive production of energy crops may also lead to the depletion of for instance carbon and nutrients in the soil.

l. Non-Renewable Matter-Bound Exergy Outputs (NMBEO_ENER)

Matter included in solid wastes exported from a system or lost by point-source pollution belongs to this group. Such losses are more easily controlled and reduced than another member of the group that can best be described as material dissipation, that is diffuse losses of finite materials that end up in the environment at concentrations similar to that of Earth's crust and considerably lower than in ores.

3.6.2 Public Societal Infrastructure and Methodology

This sector is highly diverse as it contains all structures and activities involved in establishing a society in a given region. This spans from an institutional level, that is structures and activities carried out by the municipality, to structures and activities established by and carried out the citizens. Thus, the word *public* refers more to common facilities and is used in the sense of Habermas (Habermas, 1992, original publication in German 1962).

The public sector is subdivided according to the same principles as given in section 3.6.1.1. The terms applied to symbolize stocks, in- and outflows, energy vs. matter, renewable vs. non-renewable within the energy sector are designated by the terms listed in the following section.

3.6.2.1 Public Infrastructure and Work Energies

According to the basics layout given in Chapter 2, the energies of the public sector are divided into energy- and matter-bound energies and further divided into sustainable vs. non-sustainable components.

The Stocks:

Renewable Energy-Bound Exergy Stocks—REBES_PUBL
Non-Renewable Energy-Bound Exergy Stocks—NEBES_PUBL
Renewable Matter-Bound Exergy Stocks—RMBES_PUBL
Non-Renewable Matter-Bound Exergy Stocks—NMBES_PUBL

The Inflows:

Renewable Energy-Bound Exergy Inputs—REBEI_PUBL
Non-Renewable Energy-Bound Exergy Inputs—NEBEI_PUBL
Renewable Matter-Bound Exergy Inputs—RMBEI_PUBL
Non-Renewable Matter-Bound Exergy Inputs—NMBEI_PUBL

The Outflows:

Renewable Energy-Bound Exergy Outputs—REBEO_PUBL
Non-Renewable Energy-Bound Exergy Outputs—NEBEO_PUBL
Renewable Matter-Bound Exergy Outputs—RMBEO_PUBL
Non-Renewable Matter-Bound Exergy Outputs—NMBEO_PUBL

In the following sections, an attempt is made to divide the structures and activities belonging to the public sector into the three respective groups, with a discussion as to whether they play an important role or whether they can be identified or defined at all.

3.6.2.2 Stocks Work Energy of Public Infrastructure

This section deals mainly with the amount of work energy contained in the infrastructure of the municipality.

a. Renewable Energy-Bound Exergy Stocks (REBES_PUBL)

Currently no permanent storage facilities exist (see Section 3.6.1.2).

b. Non-Renewable Energy-Bound Exergy Stocks (NEBES_PUBL)

The amount of non-renewable work energy stored is limited to the amounts of coal, oil and so on in the storage facilities in buildings owned by the municipality.

c. Renewable Matter-Bound Exergy Stocks (RMBES_PUBL)

The amount of work energy in buildings owned by the municipality or by citizens may be grouped here.

The stock may be calculated from information about the area of the buildings involved and their relative material composition and work energy content as discussed in Section 6 in this chapter:

$$WE_{RMBES\_PUBL} = \sum area \cdot fraction(s) \cdot WE_i \qquad (3.3)$$

d. Non-Renewable Matter-Bound Exergy Stocks (NMBES_PUBL)

The major stocks owned by the municipality and bound up in matter is the asphalt/bitumen products of tarmac roads. This has been estimated by the length of paved roads from the GIS maps of the municipality multiplied by an assumed average width of the roads.

No other significant stocks in this category have been identified, although minor storage of materials containing finite elements can occur in the storage facilities of the municipality; this is not likely to play a minor role.

3.6.2.3 Inflows of Work Energy to Public Infrastructure

A variety of inflows are needed to supply the public infrastructure with the work energies used to run the society. These may range from the erection of new structures or replacement of existing ones to the energies needed to run equipment, such as vehicles.

e. Renewable Energy-Bound Exergy Inputs (REBEI_PUBL)

If the energy supplied as electricity is mainly produced by wind turbines, the amount supplied is recorded here. Otherwise, it should be placed in the next grouping.

f. Non-Renewable Energy-Bound Exergy Inputs (NEBEI_PUBL)

Energy used by public or common ownership facilities for instance for running their heating systems by means of fossil fuels should be recorded here.

Likewise, vehicles owned by the municipality and running on gasoline and diesel fall in this group.

g. Renewable Matter-Bound Exergy Inputs (RMBEI_PUBL)

In principle, food entering the canteens and kitchens of retirement homes, kindergartens, schools and hospitals should be entered here.

The overall consumption of food by citizens was considered under private households so this flow is likely to have been included and accounted for there.

h. Non-Renewable Matter-Bound Exergy Inputs (NMBEI_PUBL)

All materials entering the public infrastructure for new construction or repairs should be entered here. As construction of new structures in the society is a very discontinuous process, the importance of this input is highly variable between years and is difficult to estimate.

Estimation of the replacement of existing structures can be based on rough estimates; for instance, for repair or remodelling of buildings and for bitumen for roads, a replacement rate of 1% is assumed.

3.6.2.4 Outflows of Work Energy from Public Infrastructure

Many components leave our society. Many of them are easy to comprehend because they are visible, for example solid wastes from our everyday consumption or the outflow of a wastewater treatment plant. Many others are not visible: this includes all the energy dissipation occurring in all our activities. Every time a unit of work energy is used it is usually also lost to our society.

i. Renewable Energy-Bound Exergy Outputs (REBEO_PUBL)

Work energy of this kind may be exported from the system if the energy production exceeds the local consumption. The values used in this study are all derived from the Samsø energy budget elaborated by PlanEnergi.

j. Non-Renewable Energy-Bound Exergy Outputs (NEBEO_PUBL)

This will depend on the local production and export of coal, oil and gas, but the major part will be the amount of work energy broken down in the various processes.

It seems a valid assumption that all inputs of work energies are also reflected in a downgrading of the work energy content to approximately zero. Thus, many of the import values taken from PlanEnergi may also be taken as outputs of this kind.

k. Renewable Matter-Bound Exergy Outputs (RMBEO_PUBL)

Exported renewable raw materials under municipal or other public ownership would belong to this group. In the Samsø study, only a part of the waste export—the organic fraction—would be considered here.

l. Non-Renewable Matter-Bound Exergy Outputs (NMBEO_PUBL)

The rest of the waste exported, burned or deposited and consisting mostly of non-renewable resources should be recorded here.

3.6.3 Private Households and Methodology

The people on the island have their own private lives—a sphere that partly builds on the infrastructure and facilities made available by the public authorities but also demands the establishment of additional structures such as housing facilities and the equipment needed to meet the requirements of everyday life.

The facilities have lifetimes that vary considerably, ranging from the lifetimes of houses, cars or other vehicles, kitchens with domestic appliances, electronic equipment, food and "short-lived consumable goods".

Unfortunately, many of the possible stocks and flows are difficult to account for since no data are readily available, and if they are, the consumption estimates have been accounted for in terms of money.

3.6.3.1 Energy and Private Households

The private household sector is subdivided using the same principles as given in Section 3.6.1.1. The terms applied to symbolize stocks, in- and outflows, energy vs. matter, renewables vs. non-renewables within the energy sector are designated as follows:

The Stocks:

Renewable Energy-Bound Exergy Stocks—REBES_PRIV

Non-Renewable Energy-Bound Exergy Stocks—NEBES_PRIV

Renewable Matter-Bound Exergy Stocks—RMBES_PRIV

Non-Renewable Matter-Bound Exergy Stocks—NMBES_PRIV

The Inflows:

Renewable Energy-Bound Exergy Inputs—REBEI_PRIV

Non-Renewable Energy-Bound Exergy Inputs—NEBEI_PRIV

Renewable Matter-Bound Exergy Inputs—RMBEI_PRIV

Non-Renewable Matter-Bound Exergy Inputs—NMBEI_PRIV

The Outflows:

Renewable Energy-Bound Exergy Outputs—REBEO_PRIV

Non-Renewable Energy-Bound Exergy Outputs—NEBEO_PRIV

Renewable Matter-Bound Exergy Outputs—RMBEO_PRIV

Non-Renewable Matter-Bound Exergy Outputs—NMBEO_PRIV

3.6.3.2 Stocks of Work Energy in Private Households

As mentioned earlier, the stocks of private households are highly variable, and most likely the stock proper is dominated by the housing units themselves.

a. Renewable Energy-Bound Exergy Stocks (REBES_PRIV)

These are treated similarly to those in Section 3.6.2.2.

b. Non-Renewable Energy-Bound Exergy Stocks (NEBES_PRIV)

Houses using fossil-fuel-based systems for heating will possess such a stock that may vary a lot over time. Probably when considered over an entire year this storage can be assumed to be held at a minimum level.

c. Renewable Matter-Bound Exergy Stocks (RMBES_PRIV)

According to the values given earlier, much of the structure of the house appears to be renewable: this applies to the stones, bricks, cement and wood and so on. Here we ignore the fact that the production of cement is rather expensive in terms of work energy, taking the view that the material—in principle—is recyclable.

The figure may be calculated as for the public facilities (see Section 3.3.6.2):

$$WE_{RMBES\_PRIV} = \sum area_{priv} \cdot fraction_i \cdot WEdensity_i , \qquad (3.4)$$

where the sector is composed of i types of buildings, with each building type occupying a standardized area and being composed of a standardized mixture of compounds (see Sections 3.2–3.4).

d. Non-Renewable Matter-Bound Exergy Stocks (NMBES_PRIV)

The products entering the house in the form of electrical machinery and other electronic facilities are recorded here. Again, the time scale differs considerably, as some fractions exist on the time-scale of the house while others have a turn-over time much more dependent on the arrival of new technology: radio and television, PCs, mobile phones, tools and kitchen equipment, to mention but a few major items.

Calculation of many of the preceding fractions is limited by lack of data or data that are available only in the form of monetary consumption.

3.6.3.3 Inflows of Work Energy to Private Households

Inflows are needed in order to sustain everyday life—but some inflows are more important than others, food and clothes for example, in addition to the housing facilities (already covered in the previous section).

e. Renewable Energy-Bound Exergy Inputs (REBEI_PRIV)

The electricity consumed, when provided by wind turbines (municipal or private), photovoltaics or photo-thermal installations must be considered in this group.

f. Non-Renewable Energy-Bound Exergy Inputs (NEBEI_PRIV)

The amount of work energy necessary to heat houses by use of coal, oil and gas (except biogas) is accounted for here.

g. Renewable Matter-Bound Exergy Inputs (RMBEI_PRIV)

Maintenance is an issue here and the need for this input may be estimated as earlier by multiplying a replacement or repair rate with the work energy of the housing facilities.

One of the major items is food and clothes consumed as they are all mostly derived more or less directly from biomass.

h. Non-Renewable Matter-Bound Exergy Inputs (NMBEI_PRIV)

Most other inputs to the stocks mentioned earlier are of a non-sustainable kind, including cars and metals in kitchen utilities or in electronic equipment.

The increasing consumption and the gradual increase in material stocks poses an emerging problem as items stay here until they eventually leave as waste to be recovered or deposited.

3.6.3.4 Outflows of Work Energy from Private Households

Most of the energy or matter consumed from private households ends up as dissipations (lost work energy) or solid wastes.

i. Renewable Energy-Bound Exergy Outputs (REBEO_PRIV)

In principle, houses producing electricity in excess of their own consumption should be accounted for in this group.

So far, no data for this have been identified, and it is assumed that in this case study such energy production (from photovoltaics or small "domestic" wind turbines) was of little or no importance in 2011.

j. Non-Renewable Energy-Bound Exergy Outputs (NEBEO_PRIV)

None identified in this case study.

k. Renewable Matter-Bound Exergy Outputs (RMBEO_PRIV)

If organic materials are sorted and exported in a separate fraction to be used, for instance, in the production of biofuels or composting this activity would clearly belong to this group.

l. Non-Renewable Matter-Bound Exergy Outputs (NMBEO_PRIV)

While most wastes are sorted on the island some fractions are exported, mainly to other municipalities and waste treatment facilities in Jutland.

3.6.4 Agricultural Sector—Methodology

This chapter includes considerations on how to integrate the various proposals for systematic and systemic approaches to the analysis of agricultural ecosystems, including forestry and fisheries.

In consequence, it has been found necessary to divide this sector into four sub-sectors so as to simplify the analysis of each sub-sector. Meanwhile, the first two sectors can sometimes be closely interlinked. Crops provide fodder for livestock; livestock potentially provide nutrients that are returned to the fields. Forestry may be viewed as a special case of agriculture; this is particularly obvious in the case of areas with Christmas trees. Historically, fisheries have had a close connection with agriculture, but this has loosened over recent decades.

Only a few classical attempts have been made to analyze agriculture from a systemic viewpoint based on an analysis of its energetic flows. These are mainly found in the works of the American researcher David Pimentel, presented in several books (e.g. Pimentel, 1980; Pimentel and Pimentel, 1996).

Only a very few synoptic studies seem to have been carried out on Danish agriculture and related activities (Dalgaard et al., 2002). One other source for similar agricultural systems has been identified (Hovelius, 1997). Much specific knowledge can be obtained from the advisory centres established by the government and agricultural organizations. What remains is to put the picture together which can be done from various instructive books made for the farmers from Danish advisory function or farmers organizations (Ancker et al., 2011; Jørgensen, 2008).

Hence it seems convenient to divide the agro-ecosystems into four parts:

- Agricultural crops (see Section 3.6.4.1)
- Livestock (see Section 3.6.4.5)
- Forestry (see Section 3.6.4.9)
- Fisheries (see Section 3.6.4.10)

Each of the four sub-sectors will receive special attention in the following sections, referred to earlier.

The public sector is divided according to the same principles as given in Section 3.6.1.1. The terms applied to symbolize stocks, in- and outflows, energy vs. matter, renewable vs. non-renewable within the energy sector are designated by the terms found in Section 3.6.4.1 for agricultural crops, Section 3.6.4.5 for livestock, Section 3.6.4.9 for forestry and Section 3.6.4.10 for fisheries. For the latter case of fisheries, only the terminology is given as the importance in this specific case is considered to be minor.

3.6.4.1 Work Energy and Agricultural Crops

This section deals with the work energy relationships in agricultural crops, spanning a huge variety of production. According to crop-type groupings found at the Danish Ministry of Food, Agriculture and Fisheries, there are approximately 250 different types altogether. The true number is likely to be even higher as, for example, the crops grouped under the type "cereals" are 17 in number, but these already cover at least 25 specific crops which might possess a variety of characteristics. When it comes to specific crops such as perennials, some of these will also exhibit varying outcomes over the years of cultivation, which may lead to a further subdivision of the crops.

At the same time, when dealing with work energy many parameters will vary according to farming practices (e.g. traditional vs. organic), crop age, soil treatment and use of chemicals, a situation that clearly increases the complexity of carrying out a precise work energy estimate.

Meteorological conditions will also influence the resulting analysis. To avoid this increase in complication for this study, which concentrates more on the development of techniques and tools, we have tried to circumvent this problem by using average productions as much as possible.

The main terminology used is the following:

The Stocks:

Renewable Energy-Bound Exergy Stocks—REBES_AGRI

Non-Renewable Energy-Bound Exergy Stocks—NEBES_AGRI

Renewable Matter-Bound Exergy Stocks—RMBES_AGRI

Non-Renewable Matter-Bound Exergy Stocks—NMBES_AGRI

For Inputs:

Renewable Energy-Bound Exergy Inputs—REBEI_AGRI

Non-Renewable Energy-Bound Exergy Inputs—NEBEI_AGRI

Renewable Matter-Bound Exergy Inputs—RMBEI_AGRI

Non-Renewable Matter-Bound Exergy Inputs—NMBEI_AGRI

For Outputs:

Renewable Energy-Bound Exergy Outputs—REBEO_AGRI

Non-Renewable Energy-Bound Exergy Outputs—NEBEO_AGRI

Renewable Matter-Bound Exergy Outputs—RMBEO_AGRI

Non-Renewable Matter-Bound Exergy Outputs—NMBEO_AGRI

3.6.4.2 Work Energy in Stocks of Agricultural Crops

Stocks in the agricultural crop system will vary so much that the treatment here cannot be exhaustive, so this section attempts to identify the major elements to be included in such an accounting system or inventory. Most stocks are temporary in nature, varying with the yearly production cycle and probably kept at a minimum.

a. Renewable Energy-Bound Exergy Stocks—in Crops (REBES_AGRI_CROP)

Same as Section 3.6.3.

b. Non-Renewable Energy-Bound Exergy Stocks—in Crops
(NEBES_AGRI_CROP)

Some individual storage facilities may be needed to stock fuel for equipment, such as tractors that are needed in most soil-handling and harvesting processes.

c. Renewable Matter-Bound Exergy Stocks—in Agricultural Crops
(RMBES_AGRI_CROP)

Some temporary stocks may exist over the year—mainly organic components, such as the seeds for the crops of next season, and hay and straw, either for fodder and bedding material, or to be used for heating. Fertilizers such as sludge and slurry must, in fact, be stored during certain periods to be used at the right time during the production of crops.

Parts of the buildings and many fences will be constructed from renewable or even organic materials (wood) and therefore belong to this group.

d. Non-Renewable Matter-Bound Exergy Stocks—in Crops
(NMBES_AGRI_CROP))

There may be temporary stocks of chemicals such as artificial fertilizers and pesticides, but these are also assumed to be kept at a minimum level.

Another part of buildings and fences constructed of non-renewable materials—mostly metals—fall into this group, as does equipment such as tractors and soil-handling devices.

3.6.4.3 Inflows to Agricultural Crops

According to the basic layout given in Chapter 2 the inflows to the agricultural sector may be divided into energy- and matter-bound energies and further subdivided into sustainable vs. non-sustainable components.

a. Renewable Energy-Bound Exergy Inputs—to Crops (REBEI_AGRI_CROP)

This pool comprises only the energy inputs derived from sustainable energy sources, such as heat from solar panels and electricity from either photovoltaics or wind turbines.

b. Non-Renewable Energy-Bound Exergy Inputs—to Crops
(NEBEI_AGRI_CROP)

Similarly, these flows represent the energy inputs originating from fossil fuels.

In agriculture, the major part of the energy consumption is diesel fuel used for the many types of crop handling. According to Dalgaard et al. (2002), this consumption may be roughly divided into activities related to the following activities:

- Soil handling and sowing of crops
- Handling of fertilizers and pesticides
- Plant protection
- Harvesting, threshing and bundling, and so on
- Transport and handling

Most of the values in the preceding report are given in L ha^{-1}, and therefore, only the area of a specific crop is needed in order to determine the value applying to a crop grown in a given area. In cases where other units are used, the values given in this study have been converted into L ha^{-1} by adopting typical values for the specific crops (Dalgaard et al., 2002)

Soil handling and sowing affect crop growth and consume energy and matter. The number of interventions by the farmer to prepare the soil for growing a specific crop varies greatly depending on the crop and the soil type.

The number of recommended applications of fertilizers can usually be obtained from the cultivation manuals of the specific crops. In the case of Denmark, the advisory centre for Danish Agriculture has a good deal of information on most of the commonly grown crops and the related activities. For other crops, data could only be found for organic cultivation and for more intensive types of farming, such as horticulture and greenhouse cultivation.

The energy consumption in plant protection activities is mainly related to the number of applications of herbicides, insecticides and fungicides, among others, but other treatments such as mechanical weeding may also be included.

The energy investment connected with preparing the soil, further soil manipulation during the growth season and finally when harvesting the various crops also differs considerably. Above all, the water content determines whether a certain crop needs to be dried further after harvesting, and this can represent an important energy investment in some cases.

The energy cost of general transport and other handling, for instance getting to and from the fields and transport connected with crop production in general, is considered to be relatively unimportant in this case.

c. Renewable Matter-Bound Exergy Inputs—to Crops (RMBEI_AGRI_CROP)

Seeds:

Calculation of initial values for crops:

A general equation (from the literature) is used to calculate the amount of seed needed as inoculum or sowing amount (SA) in kg ha^{-1} (Da: *udsædsmængde*) for a certain crop:

$$SA = \frac{no\_of\_plants \cdot tgw}{sprouting\_pct}, \tag{3.5}$$

where no_of_plants is the number (density) of desired plants in the field (per areal unit), that is #n ha^{-1}; tgw is the so-called thousand grain weight, given in grams (per 1,000 grains); and sprouting_pct is the percentage of grains that successfully turn into emergent plants (this factor should be seen as a gross survival as it covers both sprouting proper and survival in the initial period of crop establishment), given as the number of successful plants per 100 grains.

As an example, the desired number of plants in crops of Faba beans is 35 to 40 plants m^{-2}. Faba beans have a tgw of 550 g, and the sprouting efficiency is given as 90%, which makes it possible to calculate a seeding amount:

$$SA_{Vicia\_faba} = \frac{37.5 \cdot 550}{90} = 229.2, \tag{3.6}$$

meaning that an initial seed amount of approximately 230 kg ha^{-1} should be used to sow an average field with this crop.

d. Non-Renewable Matter-Bound Exergy Inputs—to Crops
  (NMBEI_AGRI_CROP)

A varying amount of chemicals as fertilizers and pesticides is used during crop growth and should be considered as belonging to this group; some relevant exergy values can be taken from Hovelius (1997) or needs to be estimated either from the methods given by Szargut (2005) or by use of the previously mentioned exergy calculator.

3.6.4.4 Outflows from Agricultural Crops

According to the basic layout given in Chapter 2, the outputs from the agricultural sector may also be divided into energy- and matter-bound energies and further subdivided into sustainable vs. non-sustainable components.

Meanwhile, most of the crops leave the system as renewable and matter-bound work energies—either (1) to be used as food to humans or animals, (2) to enter the energy sector or (3) remain in the organic soil fraction. The latter is interesting for potential carbon sequestration and improvement of soil texture.

a. Renewable Energy-Bound Exergy Outputs—from Crops
  (REBEO_AGRI_CROP)

(as Section 3.6.4.3)

b. Non-Renewable Energy-Bound Exergy Outputs—from Crops
  (NEBEO_AGRI_CROP)

Most of the non-renewable work energies consumed are degraded and leave the system as dissipated (lost) work energy which must be considered totally degraded, that is no longer available to us. Hence, we consider these losses as belonging to this group. The relevant data have been derived from the energy budget developed by PlanEnergi.

c. Renewable Matter-Bound Exergy Outputs—from Crops (RMBEO_AGRI_CROP)

Most of the agricultural crops are harvested and leave the system to be sold for further processing or for direct consumption as fresh goods. Likewise, biomass sent to the energy system belongs to this group.

d. Non-Renewable Matter-Bound Exergy Outputs—from Crops (NMBEO_AGRI_CROP)

The non-renewable fraction comprises irreversible losses of chemicals as well as the (photo-)respiration from the crops (CO_2 released to the atmosphere). This part is rarely made available to other parts of the system and usually ends up at other levels of society such as ground-water or dissipations. It is therefore not interesting to the scope of this study.

3.6.4.5 Livestock and Work Energy

In a work energy context, the production of livestock seems to have its own purpose namely the production of work energy to export from the system, keeping the stock to a minimum—although stocks need to be maintained either by preserving a part of the material for breeding or by import from outside the system.

Particular for livestock is the fact that a major part of the input consists of organic material used as fodder, which, in general, should be considered a renewable input. Meanwhile, an evaluation of the sustainability of a specific region should include considerations about the proportion of this fodder that is produced inside the system as opposed to fodder produced outside the system.

The same applies to systems where cattle are raised for meat production, so-called stocker farms (Am.), where part of the replacement may be covered by internal production and as is usual where a considerable amount is going to be raised, the remainder is imported from the outside.

Other necessary and important parts of this production include the inputs of medicine used as growth enhancers or for treatment that may be either prophylactic, topical or acute.

In accordance with the earlier descriptions, outputs are generally considered to be renewable (although terminal to the individuals used in meat production). Other outputs consist of dairy products (milk, cheese) and other alimentary products such as eggs.

The acronyms applied are the same as in the preceding sections, expanded with the additional acronym LIST, standing for LIveSTock. The livestock sector is subdivided on the same principles as in Section 3.6.1.1. The terms applied to symbolize stocks, in- and outflows, energy vs. matter, renewable vs. non-renewable within the energy sector are designated as follows:

The Stocks:
Renewable Energy-Bound Exergy Stocks—REBES_LIST
Non-Renewable Energy-Bound Exergy Stocks—NEBES_LIST
Renewable Matter-Bound Exergy Stocks—RMBES_LIST
Non-Renewable Matter-Bound Exergy Stocks—NMBES_LIST

The Inputs:
Renewable Energy-Bound Exergy Inputs—REBEI_LIST
Non-Renewable Energy-Bound Exergy Inputs—NEBEI_LIST
Renewable Matter-Bound Exergy Inputs—RMBEI_LIST
Non-Renewable Matter-Bound Exergy Inputs—NMBEI_LIST

The Outputs:
Renewable Energy-Bound Exergy Outputs—REBEO_LIST
Non-Renewable Energy-Bound Exergy Outputs—NEBEO_LIST
Renewable Matter-Bound Exergy Outputs—RMBEO_LIST
Non-Renewable Matter-Bound Exergy Outputs—NMBEO_LIST

3.6.4.6 Livestock Stocks

The livestock of a farm consists of a number of animals of a wide range of sizes, digestive systems (monogastric, ruminants) and with a series of purposes from meat production to more permanent providers of, for instance, eggs, milk and related products. Because of this diversity, inputs and outputs differ considerably depending on the respective type and purpose of the animals.

a. Renewable Energy-Bound Exergy Stocks—in Livestock (REBES_LIST)
This is probably not applicable except in very hypothetical cases where one could imagine that, for instance, the methane in ruminants could be viewed as a storage of energy.

b. Non-Renewable Energy-Bound Exergy Stocks—in Livestock (NEBES_LIST)
This is probably not applicable at all as live animals must in some way be considered as a renewable resource which cannot be exhausted.

c. Renewable Matter-Bound Exergy Stocks—in Livestock (RMBES_LIST)
The number may vary a lot or may be relatively stable over the year as a consequence of the purpose of the livestock—being it breeding and

meat production or continuous production of, for instance, milk for dairy products. Stocks of this kind will probably need to be averaged over the year.

The number and preferably the weight of all types of livestock is needed. Most likely local organizations of farmers will maintain a more or less detailed inventory but here the data from the supervising authorities have been used.

Work energy of individual livestock types can be obtained by multiplying the mass in kg DW by the energy content and the corresponding beta value; that is

$$WE_{Livestock\_type} = \#animals_{ls\_type,i} \cdot AvWeight_{ls\_type,i} \cdot WE-density_{ls\_type,i}, \tag{3.7}$$

where $\#animals_{ls\_type,i}$ is the number of a certain animal type i (e.g. pigs in accordance with size), $AvWeight_{ls\_type, i}$ is the average weight of animals in the group and $WE–density_{ls\_type,i}$ is the specific work energy content (usually the same for all animals, approximately 18.7 kJ g^{-1}). An extra factor—a so-called beta value for each livestock type must be used to expand the equation if one wishes to include considerations on eco-exergy values.

d. Non-Renewable Matter-Bound Exergy Stocks—in Livestock (NMBES_LIST)

This is probably not applicable unless one prefers to account some of the equipment specifically used for the animals to this stock.

3.6.4.7 Livestock Inflows

Livestock receives inflows quite different in character depending on their purpose, but common and essential to all groups are the fodder component, the inputs necessary to ensure the viability of the stock (e.g. bedding) and usually some additional nutrients, vitamins and medicines to ensure a satisfactory health status of the animals.

e. Renewable Energy-Bound Exergy Inputs—to Livestock (REBEI_LIST)

The raising and stocking of animals do require energy inputs, usually as heat and/or light, and these are included here when the energy is supplied by renewable energy production forms. In addition, transport of livestock using biofuels could be included here.

f. Non-Renewable Energy-Bound Exergy Inputs—to Livestock (NEBEI_LIST)

As for (e), but this category covers the energy inputs derived from fossil fuels.

g. Renewable Matter-Bound Exergy Inputs—to Livestock (RMBEI_LIST)

This input to livestock is essential and is likely to be a major part of the estimate as it is closely linked to two major factors: the import of breeding material and the use of fodder to feed and raise the livestock.

Here material such as the fodder used to sustain the animals should be recorded. The renewable inflows of matter-bound work energy to livestock is mainly chemically bound in the biomass or in organic substances used for fodder and in other materials such as bedding.

For a given type of stock, the necessary figure can be estimated as

$$WE_{fodder} = fodder\_demand\_per\_ind \& year_{ls\_type,i} \cdot$$
$$\#animals_{ls\_type,i} \cdot WE_{fodder}. \tag{3.8}$$

The fodder demand in $d^{-1} y^{-1}$ may be found directly for instance for dairy animals or for stocker cattle by integrating food consumption over the period. Values for various types of livestock may be found in *Håndbog til driftsplanlægning* (HDP, 2008). Similar handbooks are probably found at agricultural universities in most countries.

Fodder consumption may be given as fodder units (Da: *foderenheder*, FE). The relation between this value and biomass varies with the material used. On average, and for convenience, one fodder unit is assumed to be equal to 1 kg DM.

In fact, a considerable amount of other renewable materials such as hay and straw may be used for instance as bedding material. In this case, an extra input is needed:

$$WE_{bedding} = bedding\_material\_per\_ind \& year_{ls\_type,i} \cdot$$
$$\#animals_{ls\_type,i} \cdot WE_{bedding}. \tag{3.9}$$

So far, no data have been found to estimate this, and the values seem to vary with the practices of individual farmers. The total input necessary per livestock type equals the sum of (fodder + bedding).

To reach the total for the region, we need to sum all the livestock types in the region under consideration.

h. Non-Renewable Matter-Bound Exergy Inputs—to Livestock (NMBEI_LIST)

Here one should include salts, micronutrients, vitamins and medicine used in the nutrition. No values are currently available for this calculation.

3.6.4.8 Livestock Outflows

i. Renewable Energy-Bound Exergy in Outputs—from Livestock (REBEO_LIST)

This point probably has no relevance today, but in the future, it might be at least qualitatively important as much effort is being put into reducing the emissions of methane for ruminant livestock and, in general, to reduce emissions of gases from livestock production.

j. Non-Renewable Energy-Bound Exergy Outputs—from Livestock
   (NEBEO_LIST)

From the livestock, a part of the digested materials is lost as dissipated heat and body respiration. For that matter, the emission of CO_2 from respiration may also be included here. The amount of work energy lost by this route may be estimated as

$$\text{WE-resp} = \text{resp-rate} \bullet \text{amount ingested.} \qquad 3.10$$

This part is seen as (work) energy permanently lost from the system. Part of the irreversible losses here may eventually be determined by differences between estimates of the assimilated food and the actual growth of the animal.

k. Renewable Matter-Bound Exergy Outputs—from Livestock (RMBEO_LIST)

Livestock leaves behind partly digested food in the form of cowpats, dung, manure, slurry and so on. Such material contains not only organic materials but also valuable nutrients. The fraction is here referred to as faeces and thus used in a broad sense.

Faecalia, together with much of the bedding materials, are left as litter (manure, slurry) that may potentially be recycled within a single farm system. The efficiency of this will vary with practices and actual materials (slurry, manure, etc.).

The amount leaving the system from the individual types of livestock may be calculated as

$$\text{WE-faeces} = \text{defecation ind}^{-1} - y^{-1} \cdot \text{number of animals of livestock type.}$$
$$3.11$$

Methane emission is not assumed to represent any significant work energy loss in this context.

l. Non-Renewable Matter-Bound Exergy Outputs—from Livestock (NMBEO_LIST)

This fraction is likely to be of minor importance as compared to the part that potentially leaves as food and/or for further processing in the industrial sector. Another question is whether further processing occurs locally or not, which is a legitimate question with respect to sustainability issues.

From a system point of view, all faeces left in the fields should in principle be seen as lost, although part is likely to be retained in the system. No attempts have been made to estimate this fraction.

In principle, one could choose to view the input of work energy in nutrients, medicine and other additives as dissipated and lost to the system.

3.6.4.9 Forestry Crops and Work Energy

This section is only presented here to complete the methodology, but as no data have been available to quantify the importance of this sub-sector we have chosen not to treat these matters at the same level of details as noted earlier.

Nevertheless, the sub-sector is not totally uninteresting to the presented case study because at least one heating system in the study relies on wood-chips from forested areas—and with possible future investment in biorefin-ery processes, this sub-system may play an important role in the future.

Currently, attempts are being made to apply method to a society where for-estry plays a more important economic role than on Samsø, namely the county of Jämtland in Sweden. For initial results, see, for instance, Skytt et al. (2019).

The forestry sub-sector is divided according to the same principles as given in Section 3.6.4.1—using the acronym FORE as an extension.

The terms applied to symbolize stocks, in- and outflows, energy vs. mat-ter, renewable vs. non-renewable within the energy sector are designated as follows:

The Stocks:

Renewable Energy-Bound Exergy Stocks—REBES_FORE

Non-Renewable Energy-Bound Exergy Stocks—NEBES_FORE

Renewable Matter-Bound Exergy Stocks—RMBES_FORE

Non-Renewable Matter-Bound Exergy Stocks—NMBES_FORE

The Inflows:

Renewable Energy-Bound Exergy Inputs—REBEI_FORE

Non-Renewable Energy-Bound Exergy Inputs—NEBEI_FORE

Renewable Matter-Bound Exergy Inputs—RMBEI_FORE

Non-Renewable Matter-Bound Exergy Inputs—NMBEI_FORE

The Outputs:

Renewable Energy-Bound Exergy Outputs—REBEO_FORE

Non-Renewable Energy-Bound Exergy Outputs—NEBEO_FORE

Renewable Matter-Bound Exergy Outputs—RMBEO_FORE

Non-Renewable Matter-Bound Exergy Outputs—NMBEO_FORE

a. Renewable Energy-Bound Exergy Stocks—(REBES_FORE)

This is probably not relevant unless the whole stock of biomass is seen as a pool potentially convertible to (bio)energy by burning (which, in fact, is equivalent to [c] that follows). Meanwhile, it must be noticed that many of the exchanges with the atmosphere—assimilation of carbon (addition of energy through photosynthesis) and release by respiration will be proportional to the total activity of the biomass.

b. Non-Renewable Energy-Bound Exergy Stocks—(NEBES_FORE)

Again, when considering only the biological components such as trees, this pool is negligible. Meanwhile, in systems where forestry is more advanced, approaching the level of almost an industrialized activity, some pools may exist representing the equipment necessary to run the activities (as in agriculture).

c. Renewable Matter-Bound Exergy Stocks—(RMBES_FORE)

As mentioned under (a), the biomass storage should probably be placed here. The role of this biomass in shaping a sustainable society should not be overlooked, although this is currently being heavily debated in connection with the latest report from the Intergovernmental Panel on Climate Change (Smith, 2018). To reach a truly sustainable level, more analysis is needed of the efforts required to create such a system. This point addresses proper planning of planting and thinning and harvesting of trees, together with appropriate planning of their uses after harvest.

d. Non-Renewable Matter-Bound Exergy Stocks—(NMBES_FORE)

As under (b), this would potentially refer to infrastructural aspects concerning machinery and tools for running the activity.

In fact, implementing the analysis on a sector such as forestry will introduce a question as to how to define the boundaries of the system, for instance whether the infrastructure actually belongs to the stocks in this section or whether it should be considered as an industrial activity, that is the biological component and societal component being split and seen as belonging to two different sectors. In some cases, this will certainly be understandable and more practical.

The inflows are concerned with a number of activities ranging from the free build-up of organic material to high-tech inputs from machines which are a necessary investment for running this activity on a larger (and more efficient?) scale.

e. Renewable Energy-Bound Exergy Inputs—REBEI_FORE

The major renewable input is occurring through solar radiation and photosynthesis which as an ecosystem service leads to a free build-up of carbon-rich material.

f. Non-Renewable Energy-Bound Exergy Inputs—(NEBEI_FORE)

This point deals with the input of fossil fuels needed for the machines and other equipment used to run the activity.

g. Renewable Matter-Bound Exergy Inputs—(RMBEI_FORE)

In this case, as in agriculture, the input will consist of seeds or more likely the saplings used for (re)plantation of new forest after harvesting of biomass by clear-cutting.

h. Non-Renewable Matter-Bound Exergy Inputs—(NMBEI_FORE)

This category deals with the inputs needed to construct machinery and other equipment.

The outputs:

i. Renewable Energy-Bound Exergy Outputs—(REBEO_FORE)

The wood delivered for various activities could all be considered to belong to this group.

j. Non-Renewable Energy Bound Exergy Outputs—(NEBEO_FORE)

Usually this output will only include the dissipated energies which at this point contain little or no exergy.

k. Renewable Matter-Bound Exergy Outputs—(RMBEO_FORE)

To this category belongs all biomass derived from the system, regardless whether it arises as a result of thinning or of harvest proper.

l. Non-Renewable Matter-Bound Exergy Outputs—(NMBEO_FORE)

This section covers the use and replacement of machinery and equipment and is thus related to the turnover time of these.

3.6.4.10 Fisheries and Work Energy

As is the case with the previous sub-sector this section is only included to complete the methodology. This is not only because no data have been available to quantify the importance of this sub-sector but simply because commercial fishery from the island of Samsø has now ceased. As with forestry (discussed earlier), we have chosen not to treat these matters at the same level of detail as in the preceding sections.

The fisheries sector is subdivided according to the same principles as given in Section 3.6.1.1. The terms applied to symbolize stocks, in- and outflows, energy vs. matter, renewable vs. non-renewable within the energy sector are designated as follows:

The Stocks:
Renewable Energy-Bound Exergy Stocks—REBES_FISH
Non-Renewable Energy-Bound Exergy Stocks—NEBES_FISH
Renewable Matter-Bound Exergy Stocks—RMBES_FISH
Non-Renewable Matter-Bound Exergy Stocks—NMBES_FISH

The Inflows:
Renewable Energy-Bound Exergy Inputs—REBEI_FISH
Non-Renewable Energy-Bound Exergy Inputs—NEBEI_FISH
Renewable Matter-Bound Exergy Inputs—RMBEI_FISH
Non-Renewable Matter-Bound Exergy Inputs—NMBEI_FISH

The Outputs:

Renewable Energy-Bound Exergy Outputs—REBEO_FISH

Non-Renewable Energy-Bound Exergy Outputs—NEBEO_FISH

Renewable Matter-Bound Exergy Outputs—RMBEO_FISH

Non-Renewable Matter-Bound Exergy Outputs—NMBEO_FISH

As no relevant case studies have been identified hitherto, this section cannot yet be discussed in further detail.

3.6.5 Industry, Trade and Commerce Sector and Methodology

This sector is probably the most problematic to analyze, simply because the variation will be so great that a systematic analysis will be almost impossible. We implicitly assume here that industry proper is responsible for the major part of the work energies involved, both in terms of structures and in uses of power and materials and subsequently the export of products.

To get a proper estimate it is necessary to take our starting point in the industries of major importance. An "easy" way to start would be to use the reports of industries already involved in monitoring systems such as statutory (or voluntary) "environmental accounting" and the like.

Trade and commerce differ from industry proper in the sense that they supply society with a variety of products composed of sub-units which are more or less already produced or provide various services that require space. These sub-sectors may have a large flow through the system, but there are rarely any energy-intensive activities associated with it.

Meanwhile, the relative importance may differ widely between societies, from municipality to municipality, from region to region and so on.

3.6.5.1 *Work Energy of the Industry, Trade and Commerce Sector*

The industrial, trade and commerce sector is divided according to the same principles as given in Section 3.6.1.1. The terms applied to symbolize stocks, in- and outflows, energy vs. matter, renewable vs. non-renewable within the energy sector are designated as follows:

In Stocks:

Renewable Energy-Bound Exergy Stocks—REBES_INDU

Non-Renewable Energy-Bound Exergy Stocks—NEBES_INDU

Renewable Matter-Bound Exergy Stocks—RMBES_INDU

Non-Renewable Matter-Bound Exergy Stocks—NMBES_INDU

The Inflows:

Renewable Energy-Bound Exergy Inputs—REBEI_INDU

Non-Renewable Energy-Bound Exergy Inputs—NEBEI_INDU

Renewable Matter-Bound Exergy Inputs—RMBEI_INDU
Non-Renewable Matter-Bound Exergy Inputs—NMBEI_INDU

The Outputs:
Renewable Energy-Bound Exergy Outputs—REBEO_INDU
Non-Renewable Energy-Bound Exergy Outputs—NEBEO_INDU
Renewable Matter-Bound Exergy Outputs—RMBEO_INDU
Non-Renewable Matter-Bound Exergy Outputs—NMBEO_INDU

3.6.5.2 Stocks of Work Energy in Industry, Trade and Commerce Sector

Depending on the character of the industry, trade and commerce, stocks are, of course, highly variable. Some areas will be dependent mainly on infrastructure and buildings, some more on equipment, some more on availability of raw materials, some on labour and so on.

Some production lines can be variable in time, but it is important to notice that it is advantageous for stocked equipment to be used continuously, without void/inactive periods. This makes the stable supply and storage of raw materials even more important.

a. Renewable Energy-Bound Exergy Stocks—in ITC Sector (REBES_INDU)
(as Sections 3.6.3 and 3.6.4)

b. Non-Renewable Energy-Bound Exergy Stocks—in ITC Sector (NEBES_INDU)
Energy-intensive activities are dependent on a continuous and stable supply of fossil fuels, and this will probably be reflected in the storage of, for instance, fossil fuels.

c. Renewable Matter-Bound Exergy Stocks—in ITC Sector (RMBES_INDU)
Depending on the type of activity, stocks of renewable materials may also be important in order to ensure the stability of production.

d. Non-Renewable Matter-Bound Exergy Stocks—in ITC Sector (NMBES_INDU)
(As in Section 3.6.4)

3.6.5.3 Inflows of Work Energy in Industry, Trade and Commerce Sector

In accordance with the preceding, a continuous supply of energy and matter is necessary for an industrial system. As a minimum, stocks need to be replenished at a rate corresponding to the consumption of the materials in question.

e. Renewable Energy-Bound Exergy Inputs—in ITC-sector (REBEI_INDU)
Electricity supplied by sustainable power production is, of course, important, but possibilities also exist that plants or factories nowadays participate

in energy production themselves, for instance by installing photovoltaic cells and/or wind turbines, the use of solar thermal panels, heat exchange with heat-demanding processes and so on.

Many of these possibilities lie in the realms of cleaner production and industrial ecology, both of which have great applicational potential.

Many parts of this sector, especially in trade and commerce, will, to a large extent, be adequately served by sustainable energy sources.

f. Non-Renewable Energy-Bound Exergy Inputs—in ITC Sector (NEBEI_INDU)

Inputs of fossil fuels are certainly of great importance to many industries, but in view of the "peak oil" problematics much effort is now being invested in replacing and substituting such technologies and processes with techniques based on renewable work energy.

g. Renewable Matter-Bound Exergy Inputs—in ITC Sector (RMBEI_INDU)

Renewable matter inputs are necessary to many industries and, in some cases, may even be responsible for the largest part of the work energies involved, for example production of wood furniture, kitchen closets and the like.

h. Non-Renewable Matter-Bound Exergy Inputs—in ITC Sector (NMBEI_INDU)

Heavy industry usually refers to activities involving huge inputs of non-renewable work energy in both fuels and materials. One example is the primary processing of raw materials from ores, for instance in the iron, aluminium and copper industries.

3.6.5.4 Outflows of Work Energy from Industry, Trade and Commerce Sector

The outputs vary with the processes carried out and partly also with the level of technology used. Many previously polluting industries have been able to decrease their discharges of pollutants and even to offset the investment in new and improved technology by the cost-savings stemming from reduced use of materials and higher power efficiency.

i. Renewable Energy-Bound Exergy Outputs—in ITC Sector (REBEO_INDU)

Some factories export work energy from renewable inputs or maybe even from part of the production system itself, such as energy companies responsible for running wind turbines, hydropower plants and the like.

j. Non-Renewable Energy-Bound Exergy Outputs—in ITC Sector (NEBEO_INDU)

At the same time all industries, factories, plants and trade and commerce activities lose much of the imported work energy as dissipated heat, that is energy with no work energy value.

k. Renewable Matter-Bound Exergy Outputs—in ITC Sector (RMBEO_INDU)

Some industries produce goods that basically contain biomass and should therefore be considered as a renewable work energy export.

For some materials such as wood, the product (e.g. furniture) mostly consists of sustainable matter.

Packaging material may be derived from biomass such as paper or similar materials such as compressed corn particles. Such material should be recorded in this group.

l. Non-Renewable Matter-Bound Exergy Outputs—in ITC Sector (NMBEO_INDU)

Much of the non-renewables in this fraction arise from materials used in production but lost from the process as material dissipation or pollution.

Many products also require a huge amount of packaging material quite different in character from the renewable ones mentioned earlier. Plastics are derived from fossil fuels, and many materials involve a variety of undesirable chemicals, some of them, for instance, displaying endocrine disrupting activity. Metal containers are also handled as non-renewable but, in many cases, can be reused which must be considered in the future collection systems.

3.6.6 Nature—Evaluation and Methodology

When applying the approaches described earlier in order to analyze our society, to also shed light on the function of nature it soon becomes clear how much the two types of systems differ.

Much discussion is likely to occur when analyzing nature from a thermodynamic point of view, but probably everyone can accept the simple conversion of biomass into its corresponding energetic values. We all know that food contains energy and that wood may serve to give us heat when burned.

Much more debate and controversy may arise when also attempting to include the value of information contained in the organisms by presenting the relatively new concept of eco-exergy.

It has not been the purpose of this project to start a discussion of this issue but, rather, to evaluate whether such a distinction is important or not. Does a simple energy analysis such as is carried out in classic ecology perform well enough, or do the new concepts of work energy and eco-exergy add extra knowledge that we should be aware of in order to protect systems better, improve management, evaluate ecosystem services and so on?

3.6.6.1 Work Energy Stock and Flows of Nature

The nature sector is subdivided according to the same principles as given in Section 3.6.1.1. The terms applied to symbolize stocks, in- and outflows, energy vs. matter, renewable vs. non-renewable within the energy sector are designated as follows:

In Stocks:
Renewable Energy-Bound Exergy Stocks—REBES_NATU
Non-Renewable Energy-Bound Exergy Stocks—NEBES_NATU
Renewable Matter-Bound Exergy Stocks—RMBES_NATU
Non-Renewable Matter-Bound Exergy Stocks—NMBES_NATU

The Inflows:
Renewable Energy-Bound Exergy Inputs—REBEI_NATU
Non-Renewable Energy-Bound Exergy Inputs—NEBEI_NATU
Renewable Matter-Bound Exergy Inputs—RMBEI_NATU
Non-Renewable Matter-Bound Exergy Inputs—NMBEI_NATU

The Outputs:
Renewable Energy-Bound Exergy Outputs—REBEO_NATU
Non-Renewable Energy-Bound Exergy Outputs—NEBEO_NATU
Renewable Matter-Bound Exergy Outputs—RMBEO_NATU
Non-Renewable Matter-Bound Exergy Outputs—NMBEO_NATU

3.6.6.2 Work Energy Stock of Nature

a. Renewable Energy-Bound Exergy Stocks (REBES_NATU)
Storage of this type is probably not relevant in nature.

b. Non-Renewable Energy-Bound Exergy Stocks (NEBES_NATU)
Not relevant for as nature always is renewable.

c. Renewable Matter Bound Exergy Stocks (RMBES_NATU)
In principle, all parts of nature consist solely of stocks of chemically bound work energy, all in the form of organisms: plants, animals and microorganisms and whatever chemical compounds they are composed of.

d. Non-Renewable Matter-Bound Exergy Stocks (NMBES_NATU)
Not relevant, but in principle, the necessary stocks in the Earth's crust of nutrients and additionally required micro-nutrients and others could be included here.

3.6.6.3 Work Energy Inflows to Nature

e. Renewable Energy-Bound Exergy Inputs (REBEI_NATU)

This is the most important issue of them all—since without the input of high-quality work energy in the form of solar radiation, nature (or, for that matter, the remaining sectors) would not exist.

f. Non-Renewable Energy-Bound Exergy Inputs (NEBEI_NATU)

In most cases, this is not relevant, although one may be able to identify cases of geochemical systems and systems that are on the border of being natural, that is lightly managed systems.

g. Renewable Matter-Bound Exergy Inputs (RMBEI_NATU)

In principle all elements taken up by autotrophic organisms in addition to products derived from photosynthesis (sugar, cellulose) are renewable. This covers the macro-nutrients necessary for building up structures or for producing the required molecules, such as ATP, amino acids and proteins.

h. Non-Renewable Matter-Bound Exergy Inputs (NMBEI_NATU)

The considerations presented under (g) implicitly refer to nature as a perfectly closed system with a perfect (100%) local recycling, made possible and driven by the solar radiation entering the system. As this is almost inherent in a definition of nature, this type of input seems hardly relevant.

3.6.6.4 Work Energy Outflows from Nature

i. Renewable Energy-Bound Exergy Outputs (REBEO_NATU)

This is probably not relevant unless one includes harvesting of oils in seeds (such as rape/canola, jatropha or moringa) that may be used almost directly as a replacement for diesel in this category.

j. Non-Renewable Energy-Bound Exergy Outputs (NEBEO_NATU)

The part of inflow materials that is used in respiratory processes belongs to this category, although it represents an inevitable loss determined by the second law of thermodynamics.

k. Renewable Matter-Bound Exergy Outputs (RMBEO_NATU)

All standing stock biomass somehow ends up in this outflow, either because the organisms die for natural reasons or because it is transferred to the next step in the food chain or web (also a natural reason). Harvested biomass could also be included here (see also [i] earlier).

An important point to be made here is that nature does not leave waste as such but reuses and recycles its resources locally. Waste is always a resource for something else. In fact, some organisms such as bacteria or fungi are specialized in getting rid of the "waste"; by these processes, nutrients are recycled within the system and CO_2 released back to the atmosphere, ready to enter the photosynthesis cycle again.

l. Non-Renewable Matter-Bound Exergy Outputs (NMBEO_NATU)

With harvested biomass, some finite materials are also removed, and as with crop agriculture, this could be of importance to the future development of the natural system under consideration. Physico-geographical forcing may also be important.

3.7 Waste Management and Methodology

Again, this chapter only serves to complete the methodology, as waste was not originally within the scope of the project. For several reasons, it was decided to include the issue in the analysis anyway. Wastes as a resource (cf. nature, discussed earlier) are receiving an increasing amount of attention, and it may therefore be valuable to systematize the flows for future use (e.g. Zhou et al., 2011). In the current study, discussions about future initiatives to ensure adequate supplies of renewable fuels for the island's ferries rapidly involved issues such as biogas or biodiesel, and it became clear that knowledge of such amounts was important for the necessary decisions (e.g. Vium, 2006; Kaysen and Petersen, 2010).

3.7.1 Waste Management and Energy

According to the basic layout given in Chapter 2, the energy sector will be divided into energy and matter-bound energies and further subdivided into sustainable vs. non-sustainable components.

The Stocks of Waste:

Renewable Energy-Bound Exergy Stocks—REBES_WAST

Non-Renewable Energy-Bound Exergy Stocks—NEBES_WAST

Renewable Matter-Bound Exergy Stocks—RMBES_WAST

Non-Renewable Matter-Bound Exergy Stocks—NMBES_WAST

The Inflows of Waste:

Renewable Energy-Bound Exergy Inputs—REBEI_WAST

Non-Renewable Energy-Bound Exergy Inputs—NEBEI_WAST

Renewable Matter-Bound Exergy Inputs—RMBEI_WAST

Non-Renewable Matter-Bound Exergy Inputs—NMBEI_WAST

The Outputs of Waste:

Renewable Energy-Bound Exergy Outputs—REBEO_WAST

Non-Renewable Energy-Bound Exergy Outputs—NEBEO_WAST

Renewable Matter-Bound Exergy Outputs—RMBEO_WAST

Non-Renewable Matter-Bound Exergy Outputs—NMBEO_WAST

3.7.1.1 *Work Energy Stocks of Wastes*

a. Renewable Energy-Bound Exergy Stocks—REBES_WAST

In many waste deposits—in particular, when the waste originally had a high content of organic matter—one will find stocks of, for instance, methane that in principle is exploitable. This is known in some major cities where waste deposits have been built up over a longer period.

b. Non-Renewable Energy-Bound Exergy Stocks—NEBES_WAST

This is not likely to be relevant on the time scale considered here.

c. Renewable Matter-Bound Exergy Stocks—RMBES_WAST

Much organic matter that is deposited as waste could be considered as belonging to this group. For reasons of sustainability, it should be removed for recycling prior to deposit or incineration of the remaining waste.

d. Non-Renewable Matter-Bound Exergy Stocks—NMBES_WAST

The remaining non-organic part of the waste actually represents a potentially valuable resource if the exergy densities of the various components are higher than in ores or the Earth's crust. Hence, it should be profitable to exploit them. This is one of the reasons for distinguishing between waste dumps and deposits. However, the materials are mixed and therefore sorting and recycling of the waste before deposition are preferable.

3.7.1.2 *Work Energy Inflows of Wastes*

e. Renewable Energy-Bound Exergy Inputs—REBEI_WAST

In principle, this is not relevant, unless organic matter is deposited which could be used directly for energy production instead.

f. Non-Renewable Energy-Bound Exergy Inputs—NEBEI_WAST

In principle, depositing of energy should not be relevant, but in principle, it is possible through the deposition of residues after use.

g. Renewable Matter-Bound Exergy Inputs—RMBEI_WAST

Much of our food waste from any sector of our society ends up in this manner. Likewise, much biomaterial derived from domestic gardening is collected by the municipality instead of being composted locally.

h. Non-Renewable Matter-Bound Exergy Inputs—NMBEI_WAST

In principle, this is probably the major part of the waste collected. Increased focus on the potential for recycling has had the effect that some pre-sorting is normally carried out in most countries. This makes it possible to preserve the work energies before they actually end up as deposits.

3.7.1.3 Work Energy Outflows of Wastes

i. Renewable Energy-Bound Exergy Outputs—REBEO_WAST

Much recycling from dumps and deposits takes place in the form of gases emitted to the atmosphere. New technologies such as membranes make it possible to collect these gases—methane or CO_2—for direct use or for conversion into usable products.

j. Non-Renewable Energy-Bound Exergy Outputs—NEBEO_WAST

In principle, this does not occur, but in many cases, energy-related substances such as residues are not screened out before deposition. The consequence is that much material which otherwise would be judged as useful, for instance, to realize a circular economy end up being incinerated.

k. Renewable Matter-Bound Exergy Outputs—RMBEO_WAST

Organic materials that are produced and exploited either as a result of pre-sorting processes or by the exploitation of materials that were formerly deposited can be included here.

l. Non-Renewable Matter-Bound Exergy Outputs—NMBEO_WAST

As for (k), but these materials will take another route in the recycling system.

As seen from the preceding, much of what happens with our wastes depends on the level of sophistication of the recirculating activities in the system under consideration. In the end, this is the result of political decisions about our waste management practices.

3.8 Indices of Work Energy Efficiency of Society and Sectors

As indicated in Chapter 2 of this report, the aim of this project is to establish some indicators of the sustainability level of a society. In this case, it is considered to be possible to use work energy as a common denominator in order to establish a series of indices that can serve as indicators of sustainability. Changes in these indices can then be followed over time in order to observe the evolutionary trend of the system—whether we are moving towards or away from sustainability.

Meanwhile, it will not make sense here to evaluate all the possible indicators but, rather, to pick out a few in order to illustrate the strength of the approach. Potential users will be free to construct whichever set of indicators that best matches the requirements.

3.8.1 Stock Indicator

This type of indicator may serve to tell how efficient the system is in building up infrastructure and at what price in work energy. This could, for example, be expressed by the ratio of stock to input:

$$WESE = WE_{stock}/WE_{input,} \qquad 3.12$$

where WESE is the work energy stock efficiency, WE_{stock} is the work energy contained in the stock (e.g. the infrastructure of the system) and WE_{input} is the work energy necessary for building up and maintaining the stock.

3.8.2 Renewability Indicator

At some point, it will be interesting and necessary to know how the initiatives that we are undertaking have worked in terms of transforming our old fossil-fuel-based society into a system totally based on renewable resources.

This indicator will show how much of the work energy needed to run (1) each individual sector or (2) our whole society—that is provided by renewable resources. Referring to the preceding, this may be expressed as the ratio between renewable resource input and non-renewable resource input, the renewable inputs calculated as

$$WERI = REBEI<sector> + RMBEI<sector>, \qquad 3.13$$

in which WERI is the work energy in renewable inputs, calculated as the sum of renewable work energy—in energy and matter—and the non-renewable resource inputs, calculated as

$$WENI = NEBEI<sector> + NMBEI<stock>, \qquad 3.14$$

which leads to the following ratio:

$$RNII = WERI/WENI, \qquad 3.15$$

where RNII is a renewable to non-renewable input index (having a range from 0 [zero] to infinity). This index may be adequate for monitoring the first stages of a transition process, but with time, it may be better to replace this with another index that expresses the percentage of renewable drivers as compared to the total—for instance the following index might be used:

$$RIEF = WERI/(WERI + RINI), \qquad 3.16$$

where RIEF is the renewable input efficiency, which will approach 1 (expressing 100% renewable-based) as we reach a society where all necessary inputs are made up of renewable resources. This is an important step to take on the road to sustainability.

3.8.3 Output/Input Efficiency

Another important aspect may also be to evaluate the ability of the system (or components of the system) to pass on valuable work energy to other systems or other components of the system.

This may be expressed in terms of the output–input efficiency; that is

$$WEOIEF = WEoutputs / WEinputs, \qquad 3.17$$

where WEOIEF is the work energy output/input efficiency calculated as the ratio between the work energy in the outputs (WEoutputs) and the work energy in the inputs (WEinputs), without considering whether the work energies are of a renewable or non-renewable character.

The preceding measures are expected to serve better as indicators than just numbers derived for the various sectors after intensive investigation of their respective work energy balances. Numbers alone tend to be normative and cannot be seen in isolation since such a view may make it impossible to recognize the real situation. Therefore, some simple aggregated measures are probably better for properly elucidating the development over time in order to see, for instance, what has resulted from the measures that have been implemented.

As indicated, further divisions and distinctions can be made, and these are discussed at the end of this book. Furthermore, much of the preceding, at present, may be evaluated as not existing and hence of no importance. Meanwhile, the stocks and flux structure of each section has been retained first of all to preserve consistency in the approach. Second, once new techniques for storing energies or slowing-down processes have been realized, some the storages and flows might eventually come into play in the future. Omitting them at this state will carry the risk that they will be overlooked in future analyses.

4

Analysis of the Energy Sector

4.1 Work Energy of Energy Sector

Since 1997, the year in which the project "Sustainable Energy Island" was initiated, an internal energy analysis of the various energy flows on the island has been carried out every second year (odd years). The task of elaborating the energy budget has been carried out for the municipality of Samsø by the company PlanEnergi. The company supplies this type of service to a number of Danish municipalities in a uniform manner. The latest available budget at the time of this analysis was from 2011 as finalization and approval of the budget take considerable time. This year was therefore used as the starting point for the development and implementation of this new type of analysis.

Unfortunately, the energy budget has not been drawn up in a totally consistent manner over time. Amongst other things, the budget was developed with a view to analyzing the importance of CO_2 emissions, and as a result, its framework has been adjusted from time to time in order to satisfy current policies in this area. The changes introduced have sometimes made it more difficult to obtain an overview of the development on the island. However, preparation of the budget will be carried out in a uniform manner from now on.

More recently, some discussions have taken place about where the wind energy inputs on islands should be placed in an energy budget. Briefly, the question is whether the energy production should be credited to the island itself or to the nation as a whole. This is important, since the wind turbines established on Samsø have a capacity of approximately 35 MW for 3,700 persons; this does not represent much when divided by the total Danish population of about 5.8 million.

As the wind turbines associated with the island of Samsø are predominantly owned by the island itself, it was decided to establish a special "island energy budget" (Da.: ø-regnskab) that takes this into account. It is this version of the budget which is used as the basis for the calculations carried out here.

Although this section of the report deals with energy production and consumption in general, we also include some material-bound work energies, that is the energy contributions arising from fossil fuels and biomass. Both are here represented as the fraction entering directly into the energy budget. Of the two types, we consider the first (fossil fuels) to be non-sustainable as it is based on finite resources, whereas the latter (biomass) is considered sustainable as most of the biomass is produced on the island and in principle is renewable on a year-to-year basis. We do not here address the discussion on sustainability issues relating to greenhouse gases from different biomass practices entering the energy value chain.

The inflows to the energy sector are taken from the import and production information in the budget produced by PlanEnergi, whereas the outflow information has been derived from the consumption part of the budget. In between, there will be an inherent efficiency loss in the energy conversion along the path from the primary energy source to the energy consumed by users (i.e. in the distribution network). This happens in the case of distribution of both electricity and heat. No further attempt has been made to analyze these losses as they are considered to be inherent in the current technology; furthermore, they are dissipative losses. Stocks in the energy sector, that is the infrastructure, are considered to be part of the public sector (see Chapter 5).

4.2 Energy Import and Production

As mentioned earlier, the data for energy imports and energy production on the island are taken from information provided by PlanEnergi, with the one exception that the gross production from wind turbines has been used on the supply side, that is exports have not been deducted here, but are recorded more visibly in the calculations as real outputs.

For this analysis, the data provided have been divided into three groups:

- Imports of fossil fuels for various activities and types of consumption
- Energy provided by renewable-based technologies—wind turbines, photovoltaics and solar thermal panels
- Energy provided by biomass (usually coming from farms or forestry)

The energy flows have been converted into work energy using conversion factors found in Table 4.1. The factors reflect the devaluation of energy that occurs when transformed into another (lower) form of energy, as formulated by Brillouin (1962).

When applying the conversion factors, we observe a reduction in the value of the energy content of the energy inputs, mainly occurring in the energy fraction based on (or derived from) biomass.

The aggregated supply-side energy together with the correlated work energy content is found in Table 4.2.

TABLE 4.1

Conversion Factors for Conversion of Energy to Work Energy, That Is the Fractional Work Energy Content, Given for Various Types of Energy

Energy Type	Relative Work Energy Content
Mechanical	1
Electrical	1
Chemical	1
Nuclear	0.95
Sunlight	0.93
Hot Steam	0.6
Biomass	0.5
District heat	0.3
Heat (room temp.)	0.1
Thermal Radiation	0

Source: Values from Wall and Gong (2001), Szargut et al. (1988), Dincer and Rosen (2007).

TABLE 4.2

Supply-Side Energy and Work Energy—Imported to and/or Produced on the Island of Samsø in 2011

Input Type	Energy (TJ y^{-1})	Conversion Factor (dim.less)	Work Energy (TJ y^{-1})
Fossil fuels	284	1	284
Renewable energy production	442	1	442
Biomass	188	0.5	94
	914	–	820

Although energy is conserved, the work energy content of the imported and locally produced energy is reduced by approximately 90%, 31% of the energy and 35% of the work energy necessary to run the society is provided by non-renewable resources.

4.3 Energy Export and Consumption

As before, the data for energy exports and consumption on the island are taken from information provided by PlanEnergi (PanEnergy/Samsø Energy Academy, 2007, PlanEnergi pers. comm.), with the exception that the surplus energy from wind turbines has been recorded as exports.

TABLE 4.3

The Output—Export or Consumption of Energy and Work Energy—on the Island of Samsø in 2011

Output Type	Energy (TJ y⁻¹)	Conversion Factor (dim.less)	Work Energy (TJ y⁻¹)
Export	316	1	316
Public consumption	10	0,1	1
Private consumption	219	0,1	21.9
Agriculture consumption	19	0	0
Industrial consumption	35	0.1	3.5
Transport	66	0	0
	665	–	342.4

Besides the introduction of flow over the boundary in connection with the export of electricity, the consumption figures (mostly energy and work energy consumed by other activities) have been aggregated in accordance with the sectorial divisions and partitioning of society as described in Chapters 2 and 3.

Thus, we consider the following activities to be dominant:

- Consumption by the public sector
- Consumption by private households
- Consumption by agriculture
- Consumption by industry
- Consumption by transport

According to the budget the transport consumption corresponds to a total of 66 TJ for the island as a whole, out of which consumption by ferries accounts for 53 TJ.

The same conversion factors are used as for the inputs, with the exception that the work energy contents of the outputs are treated as if they were exhausted to room temperature or, in the case of consumption by agriculture and transport, to ambient temperature.

Combining the two estimates of inputs and outputs, that is the results given in Tables 4.2 and 4.3, in a diagram may serve to illustrate the situation of the energy sector of Samsø in 2011—see Figure 4.1.

4.4 Sustainability Indicators

The indicators established in Section 3.8 may now be calculated for the island of Samsø and for the overall energy budget from 2011.

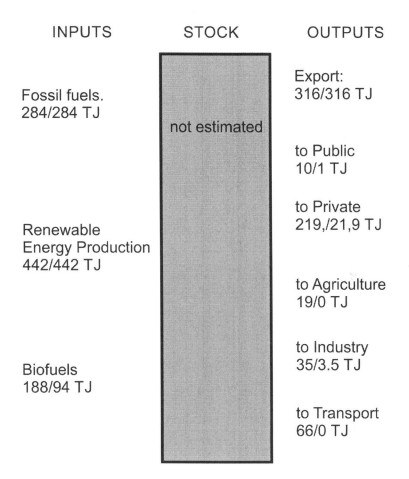

FIGURE 4.1
An aggregated diagram showing the balances of energy and work energy of the year 2011. Inputs and outputs are showing energy together with values of estimated work energy content—the values are separated by a slash. Currently no energy storages (stocks) have been identified.

4.4.1 Stock Indicator

The indicator for the overall energy budget has not been calculated as too many data are missing to justify such an exercise for the time being. However, they will mostly be present in the accounts of other sub-systems such as the public sector.

4.4.2 Renewability Indicator

If we consider the fossil-fuel-based energy as non-renewable input of work energy and the inputs from wind and solar systems together with biomass as renewable, we can calculate a renewable to non-renewable index (RNNI) for the sector:

$$RNIIenergysector = (94 + 442)/284 = 1.89$$

or a renewable efficiency RIEF of the sector:

$$RIEFenergysector = 536/(536 + 284) = 0.65.$$

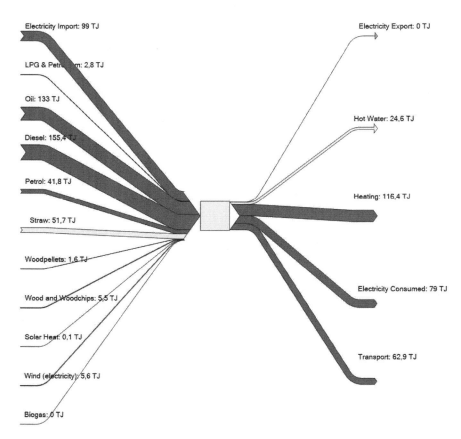

FIGURE 4.2
The flow of energy through the island of Samsø in 1997

Source:　Values taken from PlanEnergi, Samsø Energy Academy (2007).
Note:　Left: grey arrows signify non-renewable energies, light grey: biomass-related energies and black: energy inputs produced by (renewable) energy installations, primarily wind. Right: The colour black indicates dissipated energies, light grey: exported biomass-based energies and grey: exported energy derived from sustainable production. LPG = Liquified petroleum gas.

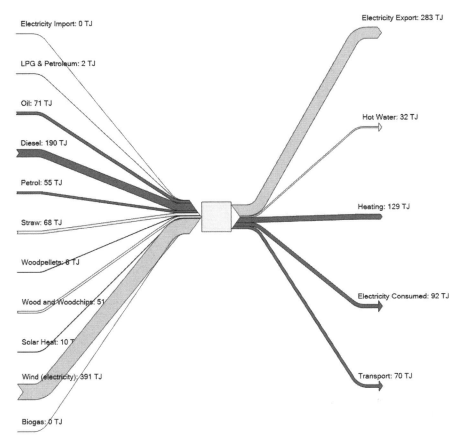

FIGURE 4.3
The flow of energy through the island of Samsø in 2011

Source: Values taken from (PlanEnergi, pers.comm., unpublished data set available from the municipality).

The total work energy input from renewable energy sources has been estimated to be 536 TJ y^{-1}, meaning that approximately two thirds of the yearly input needed to run the society is supplied by renewable work energy sources.

4.4.3 O/I Indicator

The output input efficiency based on work energy of the sector WEOIEF is

$$WEOIEFenergysector = 342.4/820 = 0.42 \ (42\%),$$

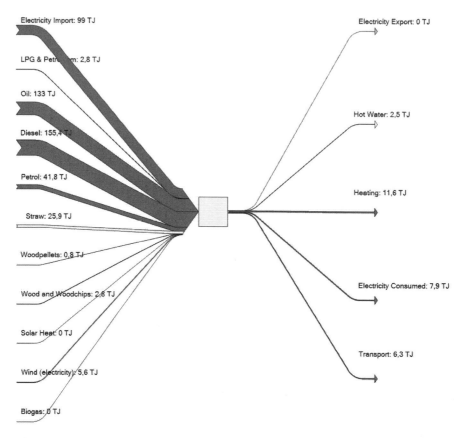

FIGURE 4.4
The flow of work energy (exergy) through the island of Samsø in 1997

Source: Values have been obtained by converting the energy values taken from PlanEnergi,
              Samsø Energy Academy, 2007.

meaning that still 58% of the input work energy is consumed (needed) by
the system.

The energy efficiency calculated as O/I ratio (and expressed as a percentage) comes to 73% whereas (as already seen earlier) the corresponding figure
in terms of work energy is 42%.

A total of 26.4 TJ of the "import" or input is consumed by the public sector,
private households and industry on the island, whereas a major work energy
fraction (71%) is exported from the island.

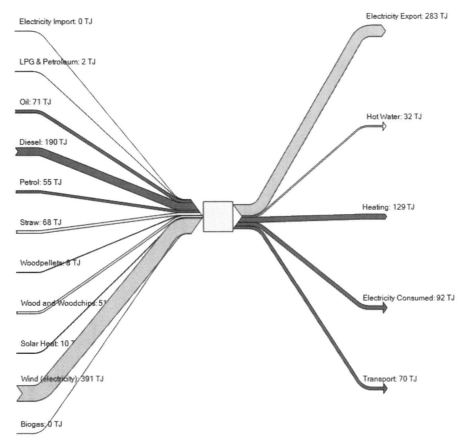

FIGURE 4.5

The flow of work energy (exergy) through the island of Samsø in 2011

Source: Values have been obtained by converting the energy values taken from PlanEnergi (pers.comm., unpublished data set available from the municipality).

4.5 Trends from 1997–2011

The methodologies used by PlanEnergy to construct the energy map of Samsø not only have changed slightly over the years, mostly around 2009 due to changes at the company itself, but also influenced by changes in legislation.

Meanwhile, a special island budget (*ø-regnskab*) has been made in which all the energy produced is also accounted for on the island. Everything else being equal this allows us to make a comparison between 1997 when the renewable energy island project was initiated and the "current" situation, represented by 2011, the latest year for which an island budget was available at the time of this study.

The changes in energy and work energy balances have been illustrated in Figures 4.2 through 4.5.

4.6 Sub-Conclusions regarding the Energy Sector

Two major observations can be made from these figures just presented. First, by the investment into wind power, the amount of energy arising from non-renewable production of electricity or heat has now decreased. Over the year, the society or municipality has become the net exporter of energy in the form of electricity which has a high work-energy value and is flexible as it may be used for many purposes in achieving sustainability. This leaves the society with many degrees of freedom to take the optimal decisions for the society on what new technologies should be implemented in the transition towards a sustainable society. Second, a tendency to see non-renewable fuels with renewable forms such as a biomass-based product is also observed. This development at present is far from being fully fledged as technologies in the area are still under development and improvement. Furthermore, some essential decisions on how to use the available biomass in an optimal manner still need to be taken. It is essential to give time to this process of negotiation amongst stakeholders to achieve the fundamental consensus on the decisions that is necessary to make this transition a similar success as the introduction of wind energy.

5

Work Energy Analysis of the Public Sector

5.1 Introduction to Public-Sector Analysis

The public sector in Danish municipalities—and probably in most other countries—depends on a certain amount of infrastructure established in order to offer a wide range of services. Thus, when attempting to estimate the importance of this infrastructure in terms of work energy, we will find it necessary to include buildings needed for the administration of the society, as well as other structures providing services to citizens at various stages of their life, from kindergartens to homes for the elderly.

In addition to this infrastructure, many functionalities become inherent in the services offered—not only the personnel required to carry them out but also, importantly, to include the equipment needed to run the infrastructure, which represents a secondary infrastructure needed in order to carry out the requested service functions.

The functionality of society today depends in many ways on a variety of services that were previously taken care of by the private sector, for instance within families. In this introduction to the work energy approach, we do not discuss the necessity or relative priorities of the various elements among the services but, rather—by making a simple inventory of activities and the related energetic costs—adopt a pragmatic approach based on the services offered and the activities normally undertaken by a society today.

Meanwhile, in the case of an island, one necessity in particular is obvious. Existence of an island society depends strongly on the transport of a variety of goods to and from the island. If not self-sufficient in energy, the island will need some sort of energy supply, often from outside providers of required fuels. At present this energy will usually be supplied from non-renewable resources, being imported as coal, oil or oil derivatives, that is fossil-based fuels.

This chapter attempts to give an overview of the major energy flows of the public sector on Samsø in terms of work energy. At the end, the actual costs in terms of work energy will be presented, and these represent objective criteria for the discussion of such services in terms of environmental sustainability.

For this sector, which includes a relatively large proportion of its infrastructure in buildings and other infrastructure, a strict categorization of flows and stocks as either sustainable or renewable in character is probably not feasible. At our present level of knowledge, it simply seems impossible to separate many of the stocks or flows into these two categories, but a preliminary attempt has been made, in order to get at a rough idea of the orders of magnitude involved.

Stocks are mainly buildings for services and administration, together with infrastructure related to transport. As we shall see for the infrastructure of other sectors, the flows into the stocks need to be estimated by assuming the rates of renewal of the various types of infrastructure.

The sector is driven by electricity and heating for buildings as well as fossil fuels to run vehicles (and ferries). This covers all the equipment needed for the society to maintain its function. At the same time, the infrastructure (mainly machinery, vehicles, buildings and roads) needs to be maintained.

Most of the outflows are consumed energies, mainly electricity and heating, but some material wastes also originate in the public system. The accounting of such outflows depends on the destiny of these wastes.

5.2 Work Energy of the Public Sector (Stock)

As mentioned earlier, this section deals mostly with the part of public infrastructure present in buildings as well as certain aspects of transport.

5.2.1 Renewable Energy-Bound Exergy Stocks (REBES_PUBL)

Since no significant storage of renewable energy (for instance as containers of Hydrogen or as compressed air) exist on Samsø at present, this point is not relevant. However, it might well be an important issue to address in the future. For the time being, this pool can only be represented by the batteries of the electric cars owned by the municipality, and up until 2011 at least, this can be taken as having no importance.

5.2.2 Non-Renewable Energy-Bound Exergy Stocks (NEBES_PUBL)

At the same time, it is assumed that fossil fuels imported to the island are consumed at the same rates as which they are imported so that they are not stored or accumulated in the system.

5.2.3 Renewable Matter-Bound Exergy Stocks (RMBES_PUBL)

The infrastructure in this section consists of the buildings owned or administered by the municipality. This pool covers the following:

- Administrative buildings proper
- Power and heating structures
- Solid waste and wastewater handling facilities
- Kindergartens and schools
- Homes for the elderly
- Social and sports facilities, libraries
- Fire brigade
- Tourist and other facilities

All in all, as shown in Figure 5.1, these stocks account for 368 TJ of work energy.

5.2.4 Non-Renewable Matter-Bound Exergy Stocks (NMBES_PUBL)

The matter in this pool is mostly considered to belong to the tarmac layer of paved roads, to which must be added the work energy contained in buildings of the sector (see Section 3.3.3).

In Chapter 3 an area of paved roads of 2 km² has been estimated. Assuming a tarmac layer of 0.1 m, a density of tarmac of 0.95 and a work energy content of 42.7 MJ kg⁻¹, the total work energy for this section can be calculated as 8113 TJ.

For the remaining part of the infrastructure, it is assumed that only a minor part of it is present as non-renewable matter such as metals, plastics and bitumen products.

The weight of the ferries (1,625 tons and 1,925 tons for the vessels *Kyholm* and *Kanhave*, respectively) adds up to 3,550 tons in total. Assuming an exergy density of about 200 MJ kg⁻¹, this corresponds to 710 TJ.

It is seen that when comparing with other stocks in the public sector, a major part of the work energy budget is contained in the bitumen part of roads.

5.3 Work Energy Inputs to the Public Sector

The data for inputs to the public sector are taken from the energy accounting analysis carried out by PlanEnergi for the municipality.

5.3.1 Renewable Energy-Bound Exergy Inputs (REBEI_PUBL)

According to the budget elaborated by PlanEnergi, the public sector of Samsø only receives a minor input of energy (as electricity) 10 TJ, which can be assumed to have the equivalent amount of work energy. As the island

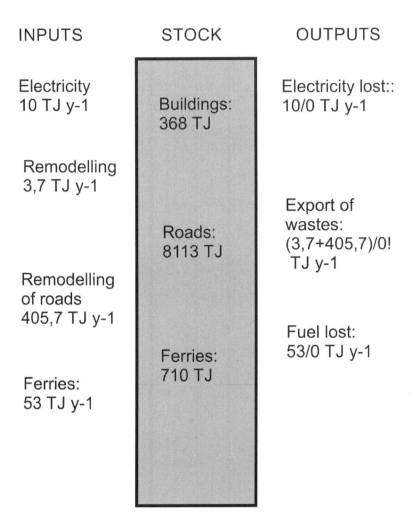

INPUTS   **STOCK**   **OUTPUTS**

Electricity   Electricity lost::
10 TJ y-1   Buildings: 10/0 TJ y-1
                   368 TJ

Remodelling
3,7 TJ y-1
                                          Export of
                   Roads: wastes:
                   8113 TJ   (3,7+405,7)/0!
                                            TJ y-1
Remodelling
of roads
405,7 TJ y-1
                                          Fuel lost:
                   Ferries: 53/0 TJ y-1
                   710 TJ
Ferries:
53 TJ y-1

FIGURE 5.1
A diagram showing the flows of work energy both in terms of the forms connected to traditional energy and material bound energies. The values have been split to show both total energy according to first law accounting together with an estimated work energy value of the respective flows. Value of work energy in stocks is indicated inside the box.

currently has a net positive production of electricity by wind turbines, this is all considered to be renewable energy.

5.3.2 Non-Renewable Energy-Bound Exergy Inputs (NEBEI_PUBL)

The major part of these inputs consists of the fossil fuels consumed by the ferries. According to PlanEnergi, their consumption amounts to about 53 TJ—containing an equivalent amount of work energy.

So far, no other non-renewable energy consumption has been identified, since no consumption of heat was indicated in the budget. In addition, most of the cars owned and operated by the municipality in 2011 ran on diesel and petrol. Their contribution to this section is likely to be minor and, in this case, is thought to be implicitly included in the budget as private consumption since no details about car ownership have been given.

5.3.3 Renewable Matter-Bound Exergy Inputs (RMBEI_PUBL)

As the building part of the infrastructure is seen consist predominantly of renewable resources, we can estimate the input necessary to maintain the structure. Assuming a lifetime of buildings to be in the order of magnitude of 100 years, we may implicitly estimate the necessary renewal rate of 0.01 or 1% y^{-1} of the stock. Thus, this input is estimated to be 3.68 TJ.

5.3.4 Non-Renewable Matter-Bound Exergy Inputs (NMBEI_PUBL)

In this case, where fossil-fuel-based derivatives like asphalt/bitumen are used for products entering the infrastructure on a limited time scale (the lifetime is likely to be considerably shorter than for buildings), these investments in road infrastructure are considered to possess work energy bound in matter of a non-renewable kind.

An estimate of the consumption of this resource may be based on the assumption of a lifetime of tarmac roads of about 20 years, leading to a need for 5% replacement per year. Following the estimate given in Section 5.2.4, this gives an absolute value of 405.65 TJ y^{-1}. The estimated replacement rate is probably too low for some roads but varies with wear and tear, which, in turn, depends on the specific usage (e.g. heavy trucks and busses cause more damage than does light traffic). For a future perspective, new technologies have been invented that permit reuse of existing material in connection with repairs. This serves to lower the work energy costs and the need for imports of non-renewable materials.

5.4 Work Energy Outputs from the Public Sector

Apart from the asphalt/bitumen products used for road surface renewals, the rest of the work energies within this sector must be considered as fully dissipated.

5.4.1 Renewable Energy Bound Exergy Outputs (REBEO_PUBL)

Following the previous argument, an equivalent of 10 TJ of electricity is dissipated; that is the work energy value in the outputs is set to 0.

5.4.2 Non-Renewable Energy-Bound Exergy Outputs (NEBEO_PUBL)

Likewise, on the basis of the inputs indicated earlier, an equivalent of 53 TJ of oil/diesel is consumed by and considered lost (dissipated) from the ferries. Again, the exergy value is set to 0.

5.4.3 Renewable Matter-Bound Exergy Outputs (RMBEO_PUBL)

A small amount of waste is estimated to originate from building materials being repaired or renewed. This may be estimated from the replacement rate—setting output equal to the input estimated from the assumed renewal rate.

5.4.4 Non-Renewable Matter-Bound Exergy Outputs (NMBEO_PUBL)

An amount equivalent to the input 811.3 TJ is assumed to leave this compartment as waste; in principle, this maintains the equivalent work energy.

5.5 Work Energy Balance of the Public Sector

An overall efficiency estimate can be calculated from the preceding, based on an output of 405.65 TJ of work energy for an input of 472.33 TJ, corresponding to an efficiency of 86%. However, these values tend to ignore the many unknown factors, and this means that although some consumptions and uses are relatively well defined, the overall efficiency estimate is associated with a relatively high uncertainty. Of course, as more detailed knowledge becomes available, the uncertainty can be reduced.

When considering that all solid wastes (and thus also the tarmac removed from paved roads) are eventually exported from the island, the efficiency of this sector tends to fall at the other end of the scale (approximately zero). As mentioned, this relationship may change with the introduction of new technologies.

The overall balance is illustrated in Figure 5.1.

5.6 Sustainability Indicators

The indicators established in Section 3.8 may be estimated for the island of Samsø based on the overall energy budget from 2011.

5.6.1 Stock Indicator

Although the amount of data is limited and incomplete, an estimated 9,191 TJ of stock must be maintained, with a total input of 472.4 TJ y^{-1} used for this purpose.

Thus, the work energy stock efficiency for the public sector is

$$WESE = WE_{stock, publicsector}/WE_{input, publicsector} = 9{,}191/472.4 = 19.5,$$

or 1,950%; that is a stock is maintained for an input of about 5% of the actual stock.

5.6.2 Renewability Indicator

If we consider the fossil-fuel-based energy as a non-renewable input of work energy, and the inputs from wind and solar systems together with biomass as renewable, we get a renewable to non-renewable index (RNNI) for the sector of

$$RNNIpublicsector = (10+3.7)/(405.7+53) = 0.03$$

or a renewable efficiency RIEF of the sector

$$RIEFpublicsector = 13.7/(13.7+458.7) = 0.03.$$

The total work energy input from renewable energy sources has been estimated as being 10 TJ y^{-1}, whereas the work energy needed for the maintenance of infrastructure (mainly road maintenance) is estimated to amount to 472.4 TJ y^{-1} in total. This leads to a low renewable efficiency.

5.6.3 O/I Indicator

The calculation of the output input efficiency based on work energy WEOIEF of the sector is highly dependent on the current waste management practices. Normally all solid waste is exported from the island (all other outputs are in the form of dissipated energies), so the value to society can be set to zero:

$$WEOIEFpublicsector = 0/472.4 = 0.$$

On this basis, the public sector comes out as having a zero (0) efficiency.

An alternative approach would be to maintain the view that the work energy of recycled building and bitumen materials is still useful, which would give an efficiency of 409.4/472.4 = 0.87—meaning that 87% is preserved, and thus, only 13% of the input work energy is consumed (needed) by the system.

5.7 Sub-Conclusions regarding the Public

From the work energy point of view, the public seems to be a cost only. This does maybe not come as a surprise when considering what type of functions and services we expect from this sector. What might be more surprising is the high costs that are related to transportation in terms of both infrastructure as such and maintenance. Not much knowledge exists in the area which could be quite critical to establishing a sustainable society. Hence, one must conclude that more information in the area needs to be gathered; that is this part of the societal system will have to receive much more attention in the future. Focus needs to be put on the necessity of remodelling efforts in particular when non-renewables are involved.

6

Work Energy and Private Sector

6.1 Work Energy and Private Households

In this chapter, we consider the role of stocks and flows in the private households, that is the place where we spend most of our time outside our working hours, the place where families meet and live. Accounting includes housing facilities as well as the equipment belonging to the daily life of households: kitchens, white goods, furniture, radio/televisions/computers and so on, which are considered by most people to be important to their lives. With increasing wealth, consumption of goods that are less basic and not essential merely to maintain life has increased. It has not been the purpose of this study to distinguish essential from non-essential needs; the aim has been to make a start based on the "typical" living conditions of most Danes today.

For the time being, some of the items can only be estimated with considerable uncertainty. For many types of goods, consumption "rates" only exist in the form of information on money spent (per capita or household) on various items. This has an impact on estimates of both stocks and flows in this sector. In fact, the inadequacy of the economic accounting approach only serves to stress the need for another way of estimating sustainability based on materials and energy consumption rather than on money.

Information about the space occupied by the housing facilities used by this sector can normally be obtained from governmental offices, the municipality or even sometimes from the internet. These data have been used as the basis for calculating the work energy in the stock of houses. For many types of consumer goods, the same uncertainty applies as mentioned earlier.

It has been much more difficult to obtain values for the flows. Again, ordinary consumption figures for electricity and fuels may be obtained from the authorities and from the energy budgets elaborated by the municipalities.

Estimates of the remaining inflows have to be derived from data held by the national authorities such as Statistics Denmark (www.dst.dk/en). Unfortunately, most of the consumption data available here are recorded in terms of their monetary values and not in amounts (e.g. kg).

For the outflows, it is possible to get information about the wastes that are handled at the places (waste dumps or deposits) established to collect municipal solid waste.

6.2 Work Energy Stocks in Private Households

As mentioned, the stocks of the private household sector are difficult to assess, with the exception of the actual infrastructure of the housing facilities. Therefore, our approach at this stage takes its starting point in the housing facilities and associated buildings.

6.2.1 Renewable Energy-Bound Exergy Stocks (REBES_PRIV)

No private houses with storage capacity for electricity have been identified, and the possible role of electric cars will be minor, even in the future. Private storage of biofuel(s) is unknown, and would probably be highly variable over time, for example seasonally dependent.

6.2.2 Non-Renewable Energy-Bound Exergy Stocks (NEBES_PRIV)

Houses heated by fossil fuels could possibly possess a stock of such non-renewable energies, but it is thought that consumption will match the imports over time so that the importance of this storage is minimal. Storage, in general, will probably tend to be minimized, although maximization (hoarding) might occur in certain situations, such as the threat of shortage due to political circumstances, an oil crisis, war in oil-producing countries and so on.

6.2.3 Renewable Matter-Bound Exergy Stocks (RMBES_PRIV)

The built-up area belonging to this sector (including the residential portion of farms) amounts to 505.670 m². This corresponds to a value of 2,187 TJ, a part that could actually have been included in the private sector.

6.2.4 Non-Renewable Matter-Bound Exergy Stocks (NMBES_PRIV)

A minor part of houses consists of a non-renewable fraction coming from finite resources such as copper for wiring. Although important for function, when looking at the values for the composition of houses this contribution to a work energy budget seems minimal and is suggested to be included in the renewable section (6.2.3)

6.3 Work Energy Inputs to Private Households

The work energy input to Samsø in 2011 was still composed of a mixture of renewable and non-renewable work energies. Although the part originating in renewable, sustainable energy production has increased significantly and has even led to a net export of electricity, a large number of private houses still rely on oil-fired boilers for heating. While renewable energy displays great potential and dominates the energy analysis, this situation shows that there is still some way for the islanders to go.

6.3.1 Renewable Energy-Bound Exergy Inputs (REBEI_PRIV)

Here we should include the private production of energy by either photovoltaics or private wind turbines. However, this was only of minor importance in 2011; in the budget of PlanEnergi, it amounts to about 1 TJ, with a similar amount of work energy. At the same time, a total of 219 TJ is consumed as power in the private sector, of which 45 TJ is supplied by "public" wind turbines.

The district heating systems deliver an amount of 72 TJ to the private households corresponding to a work energy content of about 43.2 TJ.

Work energy delivered as photo-thermal energy (8 TJ) is assumed to correspond to a value of 4.8 TJ of work energy.

6.3.2 Non-Renewable Energy-Bound Exergy Inputs (NEBEI_PRIV)

Fossil-fuel consumed by private houses corresponds to a work energy of 65 TJ and is thus comparable in size to the amount delivered via the district heating system, which is based on renewable bioenergies.

6.3.3 Renewable Matter-Bound Exergy Inputs (RMBEI_PRIV)

Part of the energy consumed by private households is delivered in the form of biomass, mainly various types of wood fuels or straw. This corresponds to 28 TJ of energy or 14 TJ of work energy.

Another contribution is represented by the amount of food that is eaten to sustain life, corresponding to an amount of 5.3 TJ (based on the average consumption of 5,000 inhabitants, that is the population including tourists).

6.3.4 Non-Renewable Matter-Bound Exergy Inputs (NMBEI_PRIV)

The inflows of other consumable goods have not been determined yet but can be assumed to be similar to the consumption by the average member of the Danish society.

The consumption of fossil fuels for transport is estimated to be 13 TJ. This is transport taking place on the island. Data from the national statistical bureau indicate that 27 cars were bought in 2011. Assuming an average weight of 1,000 kg and that much of this is steel, this corresponds to 5.4 TJ.

6.4 Work Energy Outputs from Private Households

In many ways, the private sector is the end-end station of many produced and imported products. This means that most of what is leaving the sector leaves as waste, solid or liquid, or is simply dissipated from the sector.

6.4.1 Renewable Energy-Bound Exergy Outputs (REBEO_PRIV)

This includes exported energy from private photovoltaics and wind turbines. The contribution was minor in 2011 but has increased steadily since then.

6.4.2 Non-Renewable Energy-Bound Exergy Outputs (NEBEO_PRIV)

The energies imported are all eventually converted into lost work energy (dissipation).

6.4.3 Renewable Matter-Bound Exergy Outputs (RMBEO_PRIV)

An organic/biomass fraction corresponding to 7.9 TJ is leaving as waste from private gardens.

6.4.4 Non-Renewable Matter-Bound Exergy Outputs (NMBEO_PRIV)

Domestic garbage collected every week, or large-size garbage collected less frequently, corresponds to a total amount of 136.5 TJ. At the time of the project, there was no (or only very little) sorting, and garbage is therefore necessarily seen as one pool.

6.5 Sustainability Indicators of Private Households

The indicators established in Section 3.8 may now be estimated for the island of Samsø and for the overall energy budget from 2011.

6.5.1 Stock Indicator of Private Households

The stock of work energy is estimated from the amount of infrastructure (buildings) to amount to 8,127 TJ. This stock is maintained by a total input of 202.4 TJ y^{-1}.

Thus, the work energy stock efficiency for the public sector is

$$WESE = WE_{stock,\ privatesector} / WE_{input,\ privatesector} = 8,127/202.4 = 40.2;$$

that is a highly "valued" and necessary sector is maintained at a relatively low cost, not considering maintenance and new building activities (approx. 22 TJ y^{-1}).

6.5.2 Renewability Indicators of Private Households

If we consider the fossil-fuel-based energy as a non-renewable input of work energy and the inputs from wind and solar systems together with biomass as renewable, we get a renewable to non-renewable index (RNNI) for this sector of

$$\text{RNNI}_{\text{energysector}} = (45 + 14 + 4.8 + 43.2 + 17.4)/(65 + 13) = 124.4/78 = 1.6$$

or a renewable efficiency RIEF of the sector:

$$\text{RIEF}_{\text{energysector}} = 124.4/202.4 = 61.5 = 0.65.$$

The total work energy input from renewable energy sources has been estimated to be 124,4 TJ y^{-1}, meaning that slightly more than 60% of the yearly input needed to run the private sector is made up from renewable work energy sources.

6.5.3 O/I Indicator of Private Households

The output input efficiency based on the work energy of the sector (WEOIEF) is calculated assuming that the wastes have a value of 22.1 TJ:

$$\text{WEOIEF}_{\text{energysector}} = 22.1/202.4 = 0.11,$$

meaning that only 11% of the input work energy is preserved by the system. In fact, as the wastes are exported the value is rather close to 0.

The energy efficiency calculated as O/I ratio (times 100) comes to close to 0; if wastes are reused/recycled within the island, the value can be set to the work energy value, and as seen earlier, the efficiency rises to 11%.

6.6 Sub-Conclusions regarding the Private Sector

With the introduction of wind power and a district heating network the total energy of the island is dominated by a supply of sustainable energies (see Figure 6.1). Meanwhile, a considerable amount of fossil fuels is still consumed within the sector that needs to be replaced. The sector comes out of a highly dissipative structure as most of the work energy entering the sector is lost. Only wastes may be recycled. The amount of work energy in infrastructure represents a considerable amount when assuming a replacement rate of 100 years.

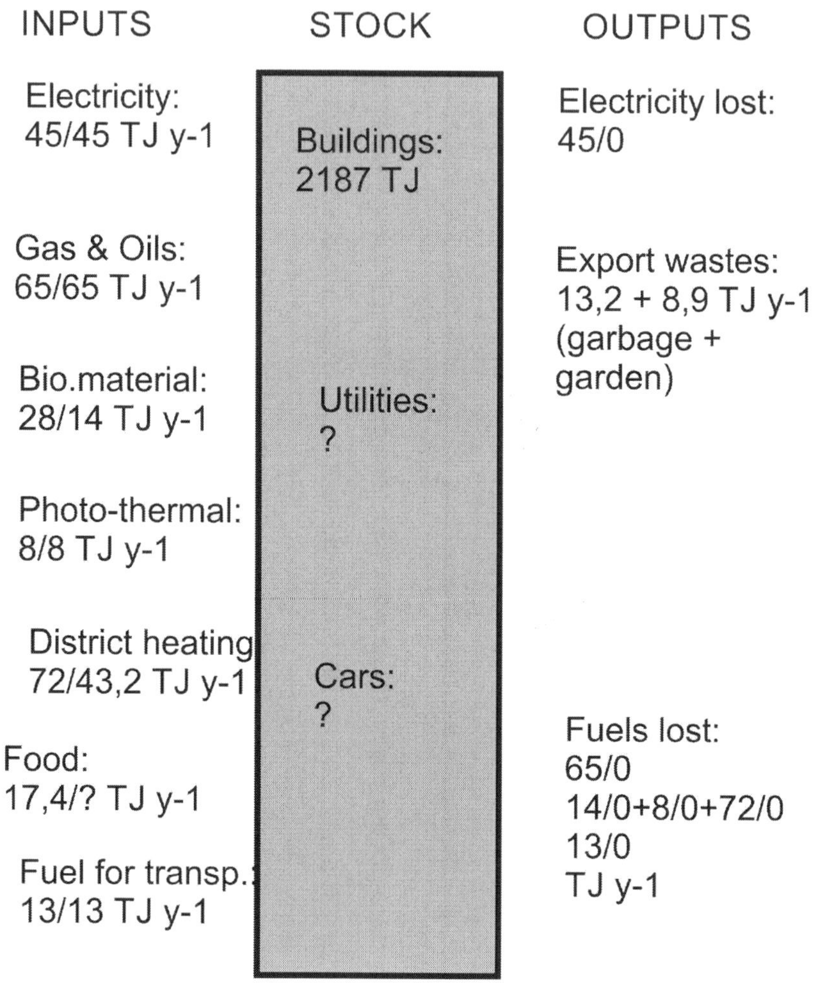

INPUTS   STOCK   OUTPUTS

INPUTS	STOCK	OUTPUTS
Electricity: 45/45 TJ y-1	Buildings: 2187 TJ	Electricity lost: 45/0
Gas & Oils: 65/65 TJ y-1		Export wastes: 13,2 + 8,9 TJ y-1 (garbage + garden)
Bio.material: 28/14 TJ y-1	Utilities: ?	
Photo-thermal: 8/8 TJ y-1		
District heating 72/43,2 TJ y-1	Cars: ?	
Food: 17,4/? TJ y-1		Fuels lost: 65/0 14/0+8/0+72/0 13/0 TJ y-1
Fuel for transp.: 13/13 TJ y-1		

FIGURE 6.1

The flows of work energy shown in the same manner as Figure. 5.1. Energy and work energy inputs show to the left are with the exception of gas, oils and transport fuels of a sustainable kind. Meanwhile, most of the work energies are lost in the structure, as only the wastes possess some values if recycling is practised.

7

Work Energy Analysis of the Agriculture, Forestry and Fisheries Sector

7.1 Introduction to Work Energy of the Agriculture and Related Sectors

This sector includes a variety of activities which are all related to the exploitation of "natural systems", be they terrestrial or aquatic. Thus, in principle, the sector spans from a more intensive use of land by agriculture via more extensive uses such as permaculture, agroforestry or "mild" management of more natural areas through grazing to managed forests that—to a greater or lesser extent—resemble nature. In this sector, we therefore attempt to include all crop and livestock production, as well as all types of forestry and fisheries.

Agriculture on the island of Samsø accounts for 75% of the land use on the island. Together with the approximately 8% of the area used for forestry of varying intensities, these types of activities involve up to 83% of the total island area.

As already mentioned, agriculture involves both the growth of actual crops together with a more traditional raising of livestock. The products of crop farming are either consumed by humans in more or less processed form, or else they enter into livestock production when used for raising animals. In the first case, the work energies are exported from the island, whereas in the second case, they stay on the island until livestock, related products or waste materials are exported.

The livestock on the island in 2011 was overwhelmingly dominated by cows and pigs, but the situation changes continuously. No slaughterhouse exists on the island, so almost all livestock is sooner or later exported from the island. All milk is also taken to dairies on the mainland for processing, with the exception of a minor quantity of "organic" milk which, during the time of the project, was used locally for farm cheese production.

On most of the island, the forestry seems to be of the low-intensity kind. That is, stands may well be artificial, and real plantations do exist, but most forest areas on the island have the character of an ecosystem with a fairly low degree of disturbance; that is they may be considered relatively undisturbed and approaching the levels of a climax plant community.

Commercial fishery is no longer practised on the island, for several reasons. The main reason is that fisheries are not economically feasible at present, as fishermen report the nearby coastal waters to be almost devoid of fish. This condition has been reported constantly during recent years and even seems to be worsening.

All in all, socio-economically speaking, the island is very much a "naturedriven" society; that is, it is driven by activities connected with exploitation of natural areas, with highly intensive production and processing of crops as well as pastures. The crops on the island consist of a diverse variety of about 90 types, most of which are exported to the rest of the country—a transfer of ecosystem services from the countryside to cities.

7.2 Work Energy and Crop Production

The following section discusses the calculation of work energy stocks, the work energy (expenditures) invested in crop-growing activities—and the various outputs of work energy from crop production. Usually, discussions focus on the economic aspects of this system, but the estimates produced during this study indicate that many potential measures could be taken within the agricultural sector which would help to mitigate the greenhouse effect, for instance through the development of techniques serving to increase carbon sequestration in the systems.

In general, the main stocks are the crops that are raised during the year, usually consisting of annual crops. Perennial and permanent crops play only a minor role in Danish agriculture today. Stock development starts from almost zero (the sowing input) to the final state of finished materials ready for harvest. It is commonly believed that as a "rule-of-thumb", the aboveground biomass (AGB-biomass) is mirrored by an equivalent amount of below-ground biomass (BGB-biomass). This must be considered a crude estimate. Nevertheless, it has been used for convenience in this study in order to estimate the total dry matter and carbon fixed in the system.

A general overview of the inputs needed to establish an agricultural crop production is as follows:

- Seeds for sowing
- Fuels for soil preparation and handling, weed and pest protection and harvest
- Electricity, or heat—for drying harvested crops
- Fertilizers, artificial and natural

- Herbicides and pesticides
- Machinery, tractors, vehicles and other equipment

In this analysis, the outputs from agricultural crop production represent a far more complex situation than we normally consider. The reason for this is that our general interest and attention are now directed not only to accounting for products which are of direct socio-economic importance; in order to establish a societal system independent of non-renewable resources, we must also pay increased attention to some crop production wastes which may become important and interesting in the future.

Thus, in addition to harvested products, the non-harvested material left in the fields, together with the root parts (unless the root is actually the part harvested, as in beets) may, for example, turn out to contribute towards conserving organic material in the soil (carbon sequestration). For cereal crops in particular, the non-grain part of the above-ground biomass may serve as an input of renewable resources to energy production (straw and biofuels such as biogas and bioethanol) or simply as bedding for livestock production.

An extended set of information is required to account for all the points mentioned before. A key item for determining inputs is the thousand grain weight (see Section 7.2.2.3) for calculation of the necessary input of sowing seed. Fuel is needed for soil handling; this quantity is taken from Dalgaard et al. (2002). The production outcomes (harvest) are taken as much as possible from the average figures for crops, gathered from a wide range of sources. If such data have not been available from the authorities, other sources have been used, such as the expected outcomes given by companies, or reports on specific crops by for instance the farmers' organizations. A harvest index has been used to estimate the total Above Ground Biomass (AGB), which also permits calculation of the Below Ground Biomass (BGB), as well as the potential biomass available for other purposes such as fodder, biogas and heating, among others.

7.2.1 Work Energy of Stocks in Crop Production

The major stocks in agricultural crop production are likely to consist of the buildings and machinery necessary for crop production. The proportion of energy crops in which the harvested materials are used more or less directly as input in the energy sector could also be grouped in this section. Likewise, the geochemical fraction of the soils could be understood as an important contributor to the natural resource input.

7.2.1.1 Renewable Energy-Bound Work Energy of Stocks in Crop Production (REBES_CROP)

Stocks of energy of this kind are probably non-existent.

7.2.1.2 Non-Renewable Energy-Bound Work Energy of Stocks in Crop Production (NEBES_CROP)

The storage of fuels on the farms is assumed to be kept at a minimum, and if so, this pool should be considered close to zero and hence negligible.

7.2.1.3 Renewable Matter-Bound Work Energy of Stocks in Crop Production (RMBES_CROP)

This part of the work energy stock consists mainly of not only buildings—but also the equipment and materials used on the farms.

The major permanent work energy stock of crop production is composed of the farm buildings (residential housing excluded, since this is considered under private households in Chapter 5) and the equipment used for soil handling: tractors, tilling equipment and other gear. The work energy of buildings on Samsø belonging to this sector is estimated to be 1,568 TJ.

The work energy stock of the crop varies over the year, probably peaking around harvest time, which, for major stocks such as cereals, rape/canola and beets is autumn. So this is a highly variable indicator which over the year may span from close to zero (e.g. bare fields in winter or spring) to a peak in biomass just before harvest (usually autumn). A conservative (average) estimate such as the amount of above-ground biomass harvested gives a figure of 3,288 TJ.

The above-ground crop stock represents an eco-exergy value of 677,622 TJ (as a live crop), assuming a general beta value for crop plants of 200.

7.2.1.4 Non-Renewable Energy-Bound Work Energy of Stocks in Crop Production (NMBES_CROP)

These stocks will mainly consist of the equipment used on the farms. No data have been identified estimating the work energy of the various types of machinery. A feasible method is needed for obtaining such an estimate.

7.2.2 Work Energy Inputs to Crop Production

The work energies of inputs are partly taken after conversion of the data from PlanEnergi, using values calculated from a number of sources in the literature. As mentioned earlier, the inputs are assumed to consist of the sowing seed and the fuels needed for soil handling, among others.

7.2.2.1 Renewable Energy-Bound Work Energy of Inputs to Crop Production (REBEI_CROP)

Initially, we here consider only the renewable inputs stemming from electricity consumption. The value is taken from PlanEnergi's energy budget, according to which the agricultural sector, together with greenhouse production, consumes 19 TJ. Correcting for greenhouse production and dividing the remainder equally between the crop production and livestock sectors, we obtain a consumption of work energy for agricultural crop production of 10 TJ. The role of this is likely to increase in future as several farms have started to erect their own wind turbines the energy of which can be used for drying harvested crops.

Meanwhile, as we shall see, this leaves the system in a rather odd situation with respect to efficiency. At some point, we need to consider that a considerable amount of solar radiation input is also consumed before we can properly assess the productivity of this sector (ecosystem services).

7.2.2.2 Non-Renewable Energy-Bound Work Energy of Inputs to Crop Production (NEBEI_CROP)

This category comprises mainly the work energy stemming from fossil fuels, predominantly diesel consumption in the various soil-handling activities mentioned earlier. Recommendations for soil handling can be found in the cultivation directions issued by various authorities giving guidance to farmers on such issues. It has been assumed that farmers tend to follow these instructions, and the calculations have been based on this.

For some crops, recommendations about soil handling are very detailed. Where such detailed recommendations could not be found, it has been assumed that types of crops which resemble each other will receive similar treatment in terms of soil preparation so that the associated energy costs will also be similar.

The fuel consumption per hectare for the various types of soil-handling activities has been taken from Dalgaard et al. (2002). Assuming (as stated earlier) that farmers follow the advice given in the instructions for cultivation, an estimated diesel consumption of 30.7 TJ can be calculated for soil-handling activities.

7.2.2.3 Work Energy Input in Renewable Matter in Crop Production (RMBEI_CROP)

The material inputs necessary to grow crops (besides the farmland that is used) are the input of sowing seeds and the chemically bound energies of fertilizers and pesticides. Here the first part—the seeds—is considered to belong to the renewable matter part, whereas the other two are considered to be non-renewable as they both depend to a large extent on energy from fossil fuels for their production, or they are derived directly from finite and non-renewable natural resources (e.g. phosphate rock).

The general equation given in the literature is used to calculate the amount of seeds needed—the seeding rate can be found in Section 3.6.4.3. The amount of seed needed has been calculated for all crops with the exception of one vegetable (Jerusalem artichoke) for which no values could be found at the time. In some cases, values from organic farming guides were used as the information was only available from such sources. Again, a large variety of sources have been used, spanning from official growing instructions to the catalogues of seed companies.

A total seed quantity corresponding to approximately 1,594-ton DW of biomass has been invested to grow the crops on Samsø. This corresponds to 29.8 TJ of work energy. Since these are living seeds, the value in terms of eco-exergy will correspond to 5,960 TJ.

7.2.2.4 Work Energy Input in Non-Renewable Matter in Crop Production (NMBEI_CROP)

The non-renewable matter-bound work energies originate from the treatment of crops with artificial fertilizers, pesticides and growth enhancers. Treatment of seeds is also included.

The amounts of chemicals such as fertilizers may also be seen as representing chemical work energy. Data for the main fertilizers are reported in an inconsistent manner; that is it is not always clearly indicated how a given work energy value is derived. Also, in several cases the relevant literature states cumulative values, that is values of work energy which also include the work energy used during the production process.

Assumptions made:

Although in the cultivation instructions several other nutrients are stated to be necessary for some particular crops, here only the values for nitrogen (N), phosphorus (P) and potassium (K) have been included. Again, it is assumed that farmers follow the cultivation instructions, and the recommended amounts and respective areas of crops are used in calculations.

For the calculation of the total inputs of N, P, and K for the growth of crops on Samsø, the values of Hovelius (1997) were used. The work energy inputs in the form of fertilizers or of pesticides are 58.1 and 5.1 TJ, respectively, or 63.2 TJ in total.

To summarize, the crop sub-sector receives an input of renewable work energy of 10 TJ, non-renewable work energy of 30.7 TJ, renewable matter-bound work energy of 29.8 TJ and non-renewable matter-bound energy of 63.2 TJ (see Figure 7.1).

7.2.3 Work Energy Outputs in Matter from Crop Production

7.2.3.1 Work Energy Output in Renewable Energy from Crop Production (REBEO_CROP)

No such sources have been identified and this output was probably not relevant in 2011. A minor and very local production of biodiesel is known to exist, but it remains and is used on the island.

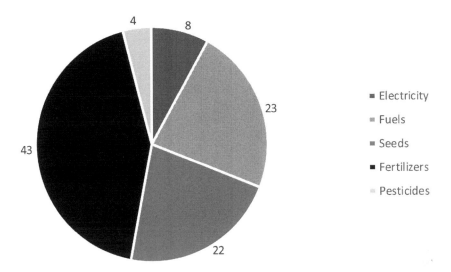

FIGURE 7.1

The work energy distribution of the various inputs to the crop production system of Samsø in 2011, which shoe the dominance of fertilizers, fuels and seeds to the production budget. These costs make up 88% of the necessary inputs of work energy.

7.2.3.2 Work Energy Output in Non-Renewable Energy from Crop Production (NEBEO_CROP)

As no energy production has been identified in the sector, no output (loss) is believed to occur, except for crop respiration, which has not been estimated. In this project, the major concern is to identify the net fluxes in the system, and respiration is implicitly accounted for in the net crop production.

7.2.3.3 Work Energy Output in Renewable Matter from Crop Production (RMBEO_CROP)

The outputs of work energies of this type are dominated by the biomasses. However, their relative importance depends greatly on what type of work energy is considered—whether it is the chemical work energy or the eco-exergy.

In addition, all energy inputs used in the sector have been lost by dissipation.

The major part of the crops produced on the island is exported either directly after being packed or indirectly after being processed at Trolleborg (see Chapter 8). Eventually, all harvested material may be considered as leaving the island; only a minor part is sold on the island to local residents or tourists. The amount of organic material produced on the island is estimated to represent a value of 817.1 TJ.

If we assume that a major part of the fodder consumed by livestock is produced on the island (or at least could potentially have been supplied by the island), we may estimate this to 375.5 TJ. Much of this may be represented as hay. If grains from cereals are used to feed cattle or pigs, careful attention should be paid to avoid double accounting, as materials used for fodder cannot also be exported.

Part of the harvested but not digestible material such as straw makes a considerable contribution to the heating systems on the island. The quantity is taken from PlanEnergi and represents an amount of 87 TJ, which may be converted (using a value of work energy density of 0.5) to represent a work energy content of 43.5 TJ.

7.2.3.4 Work Energy Output in Non-Renewable Matter from Crop Production (NMBEO_CROP)

This output includes the amount of fertilizers and pesticides exported from the sector either as run-off to streams, by breakdown, by binding to soils and/or as infiltration to ground-water.

Figure 7.2 shows the output work energies of the crop sub-sector of agriculture. The figure demonstrates a clear dominance of the work energies of exported goods.

The distribution of the remaining non-harvested biomass and the partitioning between use for fodder or heat must be considered as being quite arbitrary. First of all, the amount dedicated to fodder is the amount available,

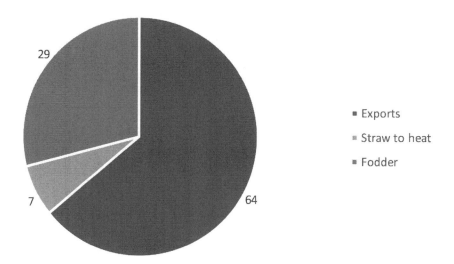

■ Exports

■ Straw to heat

■ Fodder

FIGURE 7.2

The work energy outputs of the crop sub-sector of Samsø 2011 showing the dominance of direct or indirect exports of food items from the island, as well as the large potentials for straw for heating and use as fodder.

not necessarily what is used. Second, the two possible uses are exclusive in the sense that what is used for fodder cannot be used for heat and vice versa.

The latter point serves to illustrate issues that need further consideration if the plans for a biogas plant on the island are implemented.

7.2.4 Work Energy Budget of Crop Production

The complete budget of work energies involved in the agricultural crop system may be summarized and illustrated as in Figure 7.2.

What immediately strikes the eye is that the dominant part of the biomass produced is represented by the work energy accumulated in the crop stocks. At the same time, when neglecting the input from primary production originating in solar radiation, the exports from the system exceed the inputs by a factor of 10. In principle, this could be interpreted as a highly efficient system where much arises from nothing.

This is explained by the fact that in this calculation, we have ignored the inputs from solar radiation. When viewed conservatively, this input must at least correspond to the amount of above-ground harvested biomass. This leads to an output of 1,279.6 TJ from a minimum input of 3,358 TJ—or a maximum efficiency of the system of about 38%. A recalculation including the below-ground biomass would reduce the efficiency to about half this (19%). When considering that we have dealt with net fluxes only, this is considered to be the absolute maximum efficiency of the system.

If we estimate the efficiency based on the associated destruction of eco-exergy occurring in the system, the efficiency will become much lower than this—about 0.2%.

7.2.5 Sustainability Indicators of Crop Production

The indicators established in Section 3.8 may now be estimated for the part of the agricultural sector involved in the production of crops on the island of Samsø in 2011 (see Figure 7.3).

7.2.5.1 Stock Indicator of Crop Production

The stock of work energy in this sector peaks around harvest time and is estimated from the amount of infrastructure in buildings (1,568 TJ) and the total biomass, above and below ground (3,388.1 TJ) which sums up to 4,956.1 TJ. This stock is maintained by a total input of 3,494.4 TJ y^{-1}.

Thus, the work energy stock efficiency for this sector is

$$\text{WESE} = \text{WE}_{stock,crop-sub-sector}/\text{WE}_{input,crop\ sub-sector} = 4{,}956.1/3{,}494.4 = 1.41,$$

indicating a sector with a relatively high cost. However, it is considered more appropriate not to take the ecosystem services from the sun into account.

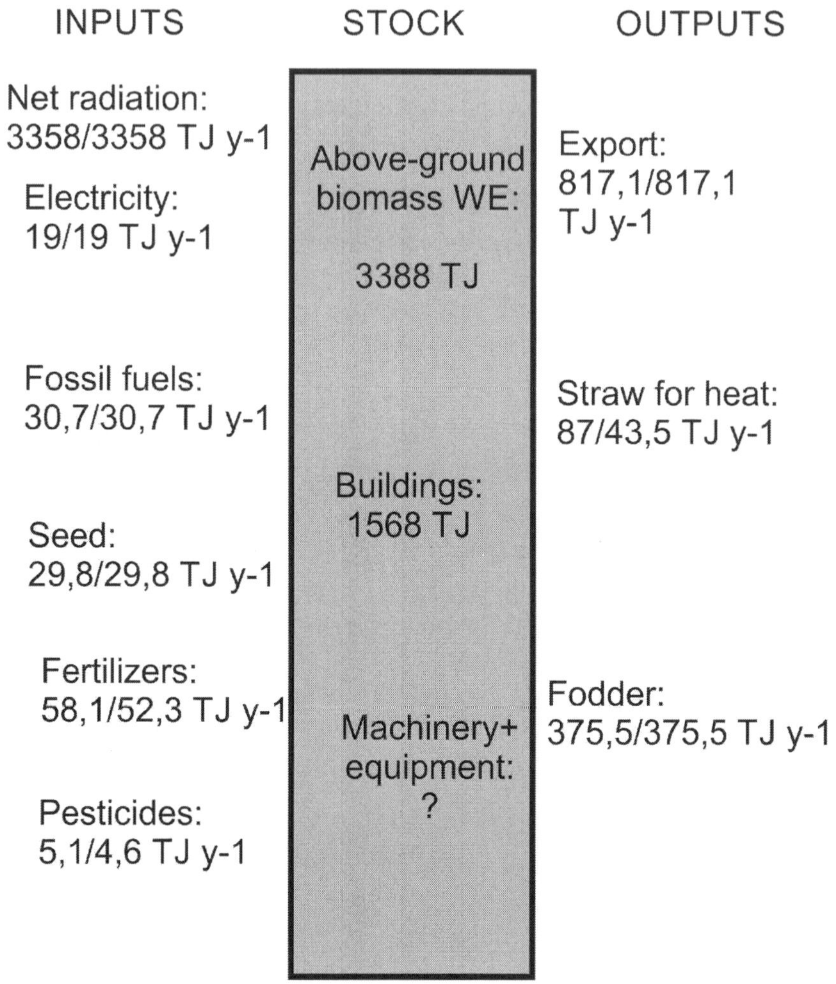

INPUTS   STOCK   OUTPUTS

Net radiation:
3358/3358 TJ y-1   Above-ground   Export:
                        biomass WE: 817,1/817,1
Electricity: TJ y-1
19/19 TJ y-1   3388 TJ

Fossil fuels: Straw for heat:
30,7/30,7 TJ y-1   87/43,5 TJ y-1

                        Buildings:
Seed: 1568 TJ
29,8/29,8 TJ y-1

Fertilizers: Fodder:
58,1/52,3 TJ y-1   Machinery+   375,5/375,5 TJ y-1
                        equipment:
Pesticides: ?
5,1/4,6 TJ y-1

FIGURE 7.3
The work energy flows involved in the production of crops on the island of Samsø in 2011. In a similar manner as earlier chapters, a slash is separating the energy and estimated work energy values. For the major living biomass components—seeds used and amount of crop produced— a value of the natural work energy including information (NWEI = eco-exergy) is given. This illustrates the bias introduced to the evaluation if this indicator is used.

This would lead to a stock value of 4,956.1 for a reduced cost of 136.4, giving a WESE figure of 36.3.

7.2.5.2 Renewability Indicators of Crop Production

If we consider the fossil-fuel-based energy as a non-renewable input of work energy and the inputs from wind and solar systems together with biomass as the renewable input, we get a renewable to non-renewable index (RNNI) for the sector of

RNNI energysector $= (3{,}368 + 19 + 29.8)/(30.7 + 52.3 + 4.6) = 3{,}406.8/87.6 = 38.9$

or a renewable efficiency RIEF for the sector of

$$\text{RIEF energysector} = 3{,}406.8/3{,}494.4 = 0.97.$$

The latter value shows that the input energies are dominated by the renewable/sustainable work energy part—a point that really stresses the importance of the ecosystem service delivered by solar radiation.

7.2.5.3 O/I Indicator of Crop Production

The output–input efficiency based on work energy for this sub-sector—WEOIEF—is calculated from the export of materials used for district heating and the potential amount of fodder used on the island. The calculated O/I efficiency would be

$$\text{WEOIEF}_{\text{crop sub-sector}} = 1236.1/3494{,}4 = 0.35,$$

meaning that 35% of the input work energy is exported by the system. This may be seen as a relatively low value, but considering that the crop system includes potential outputs which may be used elsewhere and preserves organic material in the soil, it is probably a value that may be taken to be positive. In the case, where all work energies leave the island this would correspond to exhaustion of the soils.

7.3 Work Energy of Livestock Production

This sub-sector considers the work energies that are used in livestock production. In the case of Samsø it is dominated by the production of cattle and pigs, resulting in exports of meat—or rather in the export of animals for slaughter—and some exports of milk from dairy cattle. In 2011, there was no processing of meat or dairy products on the island, although a minor organic cheese production was starting up around this time.

The amount of livestock is said to have decreased over recent decades. In particular, the activities in the area of pig production have declined as several farms have closed down. One very large farm in particular was considered when it was established to be the largest in Europe.

The data used in the section have been provided from the municipality and other supervising authorities and are based on stocks larger than

3 units. Although present on the island, no values for poultry production or numbers of sheep could be obtained.

The amounts of fodder needed to raise animals together with outputs (dung/slurry and respiration) have been estimated by use of values from the literature, mainly Pedersen (2011) and Rasmussen and Sylvestersen (2009).

7.3.1 Work Energy Stocks of Livestock Production

The stock of this sub-sector is dominated by the amount of livestock, which in 2011 is assumed to represent a relatively stable situation (see Figure 7.4).

7.3.1.1 Renewable Work Energies in Energy of Livestock Production (REBES_LIVE)

This type of work energy was not identified in 2011.

7.3.1.2 Non-Renewable Work Energies in Energy of Livestock Production (NEBES_LIVE)

As discussed previously, the amount of work energy held in fuels is included in this section but must be assumed to be kept at a minimum.

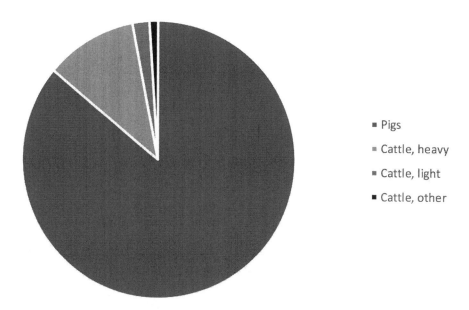

- Pigs
- Cattle, heavy
- Cattle, light
- Cattle, other

FIGURE 7.4
The work energy content of various livestock identified on Samsø 2011, showing the vast dominance of pigs which together with cattle make up close to 100% of production units on the island.

7.3.1.3 Renewable Work Energies in Matter of
Livestock Production (RMBES_LIVE)

The composition of the livestock is still dominated by the production of pigs (86%) followed by cattle of various sizes and purposes for meat and milk production (14%). These two fractions come close to representing the total livestock on the island.

While several other livestock types have been observed, it has not been possible to obtain data to quantify them properly. No large-scale poultry production exists, sheep seem to be kept for nature management only and horses play a part only in the leisure and tourism sector.

7.3.1.4 Non-Renewable Work Energies in Matter of
Livestock Production (NMBES_LIVE)

This category would include for instance stocks of internal equipment in the stables, including boxes and pens. Also, stocks of chemicals such as medicines should be accounted for here. Unfortunately, no data have been identified for either of these.

7.3.2 Work Energy Inputs of Livestock Production

The dominant input of work energy is the fodder needed for optimal production of the livestock, either as growth of the animals, that is for meat, or for maintaining optimal conditions for production and nursing of offspring, as well as the production of other outputs such as milk.

Here the detailed composition of feed materials has a large impact on the desired growth and/or production conditions, and this can potentially affect the work energy content of the respective feeds. So far, no attempts have been made to distinguish between the various fodder materials as this is considered outside the scope of this project.

It is also more likely that the real differences in the work energy values of the feeds would depend more on their background, that is the manner in which they have been produced, rather than on their particular composition.

7.3.2.1 Work Energy Input in Renewable Energy in
Livestock Production (REBEI_LIVE)

The work energy of electricity supplied to the agricultural sector (excluding greenhouse production) has been allocated equally between crop and livestock production. This consumption is thus estimated to be 9 TJ.

7.3.2.2 Work Energy Input in Non-Renewable Energy
in Livestock Production (NEBEI_LIVE)

No consumption of fossil fuels for livestock production has been registered. Although some fuel must be used for transport to and from fields this is likely to be minimal. Another contribution to transport occurs in connection with the exports of milk and animals for slaughter. This is probably also a minor figure and is probably externalized from the work energy budget of the island. Livestock also requires stables to be heated especially during wintertime, and this consumption may as experienced in northern regions like Jämtland (Sweden) represent a significant part of the energy consumed.

7.3.2.3 *Work Energy Input in Renewable Matter in Livestock Production (RMBEI_LIVE)*

The work energy contained in fodder produced on the island is considered to belong in this category.

The necessary values are derived from Pedersen (2011) and Rasmussen and Sylvestersen (2009) based on the cumulative fodder production during the production of one of the respective livestock units. The estimate arrives at a value of 375.5 TJ needed to maintain production on the island. (This is also the value used for the output from the crop sub-sector)

When compared with the crop production, this quantity of fodder and some additional materials for bedding, among others, could in theory be derived entirely from the crop production on the island itself. No identification of the role of imported fodder has been obtained so far. But fodder, like soya beans used as protein-rich fodder, is often not considered sustainable due to transport distances. Locally produced fodder is preferable.

7.3.2.4 *Work Energy Input in Non-Renewable Matter in Livestock Production (NMBEI_LIVE)*

Inputs of imported fodder and additives (e.g. soya bean cakes or medicine) would be considered to belong to this group. However, no data could be obtained for this.

7.3.3 Work Energy Outputs of Livestock Production

The major outputs from this sub-sector are its commercial products: meat or dairy products for the use of our society.

Meanwhile, the production of livestock also has well-known costs which are clearly reflected in the work energy budget. Some of the output is inevitably lost in the form of the amount of livestock exported or is lost as respiration during the life cycle of the animals.

Some of the output such as the manure and slurry is regarded as a renewable resource. Not only may it be used as an alternative to artificial fertilizers, but this resource may also play a part in future plans for biogas production.

7.3.3.1 Work Energy Output in Renewable Energy from
Livestock Production (REBEO_LIVE)

So far, no outputs are utilized for the production of biomass-related energy, and this sub-sector is therefore set to zero.

7.3.3.2 Work Energy Output in Non-Renewable Energy
from Livestock Production (NEBEO_LIVE)

The part of the work energy used up in respiration may be estimated here, calculated as part of the work energy ingested in fodder and broken down to an almost ambient temperature by this process. This is estimated to be 157.7 TJ (42% of ingestion).

7.3.3.3 Work Energy Output in Renewable Matter from
Livestock Production (RMBEO_LIVE)

One item in this sub-sector is the livestock exported for meat and other products. Assuming that most of the livestock, for example 95%, is replaced over the year this gives an estimate of approximately 36 TJ exported from the island.

The proportion of ingested material ending up as cowpats, manure or slurry is included here and is estimated to be 161.5 TJ (43% of the ingestion). No attempt has been made to determine how much ends up in fields (cowpats) or how much is potentially available from cowsheds and piggeries (manure and slurry).

7.3.3.4 Work Energy Output in Non-Renewable Matter
from Livestock Production (NMBEO_LIVE)

No losses recorded here as no data for dead animals could be found.

7.3.4 Sustainability Indicators of Livestock Production

The indicators established in Section 3.8 may now be estimated for the island of Samsø based on the overall energy budget from 2011.

7.3.4.1 Stock Indicator of Livestock Production

The stock of work energy is estimated excluding the infrastructure (buildings) which has been accounted for elsewhere. Thus, this stock consists of the animals only and is estimated to 40.41 TJ. The stock is maintained based on a total input of 386.5 TJ y^{-1}.

Thus, the work energy stock efficiency for the livestock sub-sector is

$$WESE = WE_{stock, livestock sub-sector}/WE_{input, livestock sub-sector} = 40.4/386.5 = 0.10$$

meaning that only 10% of the inputs are successfully converted into stock—thus indicating a sub-sector with low efficiency. This is partly known already from conventional analyses, but the extension to work energy seems to strengthen this point.

7.3.4.2 Renewability Indicator of Livestock Production

It is not possible to calculate the renewability ratio as this involves division by 0. This is an artefact caused by the missing data for investment of non-renewable work energies in this sector:

$$RNII_{energysector} = (9 + 375.5)/(0) = ? (\infty).$$

The renewable efficiency RIEF of the sector in consequence becomes optimal, as all inputs are in the form of renewable work energies:

$$RIEF_{energysector} = 386.5/386.5 = 1.$$

This is also a consequence of missing data—but overall, the efficiency in this context is probably high and the role of non-renewables negligible.

7.3.4.3 O/I Indicator of Livestock Production

The output–input efficiency based on work energy of the sector WEOIEF is composed of exports and outputs of faecal materials, together with the inevitable loss through respiration (with zero work energy):

$$WEOIEF_{stock, livestock sub-sector} = 197.5/386.5 = 0.51$$

efficiencies, meaning that only 51% of the input work energy is passed on by the system. The real evaluation of this figure depends on the destiny of the manure, cowpats and slurry and whether it is used within the system or not. If this part is not taken into account, the efficiency becomes low, that is about the 10% indicated earlier.

   The result accentuates the role of animal wastes as a resource in agriculture.

7.4 Work Energy of Forestry

The actual role of forestry is hard—if not impossible—to evaluate precisely on the basis of present knowledge. Meanwhile, some estimates of orders of magnitude may be assessed.

   The work energy in forest stocks amounts to 6,329.2 TJ for plants (trees) and 16.5 TJ and 2389.4 TJ in animals and plant litter, respectively. This corresponds to eco-exergies of 1,740,533 TJ, 3,296 TJ and 2,389 TJ, respectively, for the three stocks (see Chapter 9).

Some forestry activities, such as the production of conifers for Christmas trees, are often considered to be a form of agricultural production. Only a minor part—corresponding to approximately 0.2 km² out of the 8.8 km² reported as forest in the Geographical Information System—has been registered as agricultural crops.

From this, a maximum energy of 99 TJ, corresponding to a work energy of approximately 50 TJ, enters the island's energy budget according to PlanEnergi, which means that 0.8% of the work energy contained in the plant biomass of forests is consumed per year. In order for this input to be sustainable, an equal amount of energy, corresponding to the area harvested, that is the biomass removed, needs to be replaced every year.

No knowledge could be obtained about the amounts of wood, wood pellets and chips imported to the island.

7.5 Work Energy of Fishery

No professional fishery activities exist on the island any more. Although fishery still continues as a leisure activity, it is considered to be of no significance for the work energy balance of the island.

Considering that the marine area of the municipality represents an area 12 times greater than the terrestrial part, it could potentially play a larger role. At present, the marine area provides ecosystem services, which are not evaluated here.

7.6 Sub-Conclusions regarding Agriculture

The agricultural sector and the other activities included that are related to exploitation and use of nature is probably one major concern when it comes to addressing the issue of sustainability. First, many of the sub-activities are related to transports such as important materials and hence carries and implicit environmental load. This issue needs to become clear in future evaluations of the sustainability of the sector. Second, the sector has many direct energy costs and at present depend strongly on fossil fuels. Meanwhile, these issues may be repaired by the investments made in sustainable energy forms. Furthermore, the initiative must rely on more intense use of the waste materials from both crop and livestock production. The method as applied here does ignore the impact on carbon storages and the changes in biodiversity which occur as a result of the transformation of the landscape.

8

Work Energies of the Industry, Trade and Commerce Sector

8.1 Introduction to the Industry, Commerce and Trade Sector

At first look, this sector—the industry proper—appears as if it is of little or even no importance in this particular case. Clearly, this sector would become much more dominant in the analysis if Samsø had been a society with more heavy industry such as iron production, car factories and so on. However, the island possesses at least one major industrial activity, namely the processing of the many vegetables grown on the island. This is carried out by the factory "Trolleborg" on the southern-western part of the island between Tranebjerg and Kolby Kås. The factory was started in the 1960s and since then has had several constellations of ownership but always with partial local ownership involved. This section deals with the processing of goods related to agricultural crop production on the island.

Meanwhile, the activities of handling fresh vegetables and other products to be exported from the island also trigger a substantial amount of related activities which have not been accounted for in detail in the available inventories. All these activities deal with handling and storage of crops and herbs from the island on their way either to processing or to export from the island.

As seen in the following, this sector—although considered to be of low importance—relies on quite an amount of infrastructure, which represents energies of a similar order of magnitude to the other sectors in this analysis.

While the previously mentioned activity at "Trolleborg" has been established and running for more than 50 years, many other activities closely connected to agriculture and the processing of agricultural products have vanished from the island, such as slaughterhouses, dairies and the like.

Recently, a few sites with so-called niche production have appeared, including a microbrewery and a factory for organic cheese production. It is likely that more of this type of activity will appear in the future, as the inhabitants plan to re-establish both a dairy and a slaughterhouse on the island.

The various activities in this group consume a substantial amount of energy in the form of fossil fuel, that is non-renewable energy, most of the consumption by one company alone. At the same time, several other activities connected with the storage, handling and packaging of fresh goods for export must consume a certain amount of electricity, mainly for lighting and cooling, but this has not been recorded as a separate item in the energy budgets.

Although some storage material, mainly glass jars and lids, is imported to the island, a major part of the storage materials are goods produced on the island. Only a small proportion of the produced goods stem from imported materials.

Stocks in this sector, apart from the production proper, are closely related to commerce and trade activities. In particular, this not only applies to the areas occupied by the services offered to the island population throughout the year, such as grocery stores and other sales outlets, but also includes the services offered to tourists, such as hotels and restaurants. Many of these activities close down when outside the touristic season.

Only a small and highly variable proportion of the vegetables (materials) used in production is imported (various types of cucumbers). This means that a major part of the required input for maintaining production consists of the packaging materials, such as jars and lids, all representing material-bound work energies. In addition, the sector still obtains a considerable amount of its energy in the form of fossil-fuel-based work energies, as many processing procedures involve the boiling of organic materials.

Basically, all materials entering the production—that is the processed agricultural products in cans and jars—leave the island and must be considered to be an output (export) of work energies, whereas the energy inputs are lost as dissipation. An estimate of the waste outputs indicates a potentially valuable resource that is not used at the time of this analysis and therefore does not enter the calculations. Part of this waste is now being considered as potential feedstock in a future biogas plant.

8.2 Work Energy Stocks of the Industry, Commerce and Trade Sector (ICTS)

As this chapter is based more and less on one major industrial activity, the role of stocks is highly dependent on this fact. The stocks of materials to be processed, the final products and their wastes will vary strongly with season so that the only permanent stock is made up of the infrastructure, that is the industrial plant itself including buildings and machinery. For larger industries, storage may play an important role, although the individual factories tend to minimize storage for economic reasons. Such minimization is fully recognized as a target in the "leaning" of activities.

8.2.1 Renewable Energy-Bound Exergy Stocks in ICTS (REBES_INDU)

The sector to our knowledge currently contains no facilities for the storage of renewable (work) energies. As indicated, this part may play a role in the future, but the exact formulation will depend on what kind of biorefinery process is chosen.

8.2.2 Non-Renewable Energy-Bound Exergy Stocks in ICTS (NEBES_INDU)

In principle, the ICTS could experience situations where a stock of fuels was necessary or reasonable to have, but in this case—and considering the time-scale of a single year—we expect this balance to come out at zero; that is, in general, no stocks are present or relevant on this time scale. Proper planning aiming at a steady supply of energy from renewable resources should tend to eliminate this completely with time.

8.2.3 Renewable Matter-Bound Exergy Stocks in ICTS (RMBES_INDU)

The whole area occupied by ICTS infrastructure is estimated to be 100,150 m², of which the factory Trolleborg accounts for 4,958 m², corresponding to approximately 5% of the total. So, in fact, a considerable area is dedicated to the other activities grouped in this sector.

Thus, the total infrastructure of the ICTS activities on the island in terms of the work energy represented in buildings amounts to 753.6 TJ, with Trolleborg only responsible for 33.4 TJ, or 4.4% of the work energy of this infrastructure.

In principle, all building materials are renewable as long as they are properly handled, that is not disposed but reused or cycled within their life cycle. This assumes that enough energy can be supplied to run the activities connected with the recycling of these materials.

8.2.4 Non-Renewable Matter-Bound Exergy Stocks in ICTS (NMBES_INDU)

In accordance with the argument presented earlier, no attempt has been made to split the work energy in buildings into a renewable vs. a non-renewable part. However, such knowledge would be interesting for future initiatives concerning sustainability.

8.3 Work Energy Inputs of ICTS

The work energy inputs of this sector consist mainly of the materials to be processed, the storage/packaging materials and the energy needed to run the process.

8.3.1 Renewable Energy-Bound Exergy Inputs to ICTS (REBEI_INDU)

The electricity used under this heading comes almost entirely from the wind turbines on the island. The reason for this situation is that no storage facilities have yet been established to cover energy supply during periods with no wind or solar input. According to PlanEnergi, this input amounts to 6 TJ, of which 2.9 TJ is consumed by one enterprise alone.

8.3.2 Non-Renewable Energy-Bound Exergy Inputs to ICTS (NEBEI_INDU)

The amount of fossil fuels consumed by this sector has a work energy value of 18.2 TJ (PlanEnergi, special island accounting). The actual consumption may be higher, but it has not been differentiated from the rest of the energy budget for the public sector and private households.

8.3.3 Renewable Matter-Bound Exergy Inputs to ICTS (RMBEI_INDU)

Vegetables and other agricultural products are the main material inputs to this sector. They consist mainly of potatoes, beetroots, cucumbers, pumpkins, tomatoes and several other products. The total vegetable input amounts to 32.4 TJ.

Again, in principle, containers and other packaging materials are recyclable (see Section 8.2.3), and this fraction corresponds to 38.2 TJ of materials.

A considerable amount of various substances is used for conserving the foods after cooking, and the materials used here—sugar, vinegar and preservatives—represent an input of 23.1 TJ.

8.3.4 Non-Renewable Matter-Bound Exergy Inputs to ICTS (NMBEI_INDU)

Although some minor amounts of chemicals are known to enter the production and therefore should be included in the account, the quantities are considered in this context to be of no importance, that is to be 0 TJ.

8.4 Work Energy Outputs from ITCS

The outputs of this sector are dominated by the material outputs since the processed goods, together with packaging materials, are exported from the island. The work energies used during processing are considered lost (dissipated).

8.4.1 Renewable Energy-Bound Exergy Outputs (REBEO_INDU)

As no production of energy is known to exist from this sector, this part is considered not to play an important role at present, that is when discussing with the situation in 2011. Nevertheless, after significant developments in 2012 and 2013, it may be important to include this item in future studies. The potential use of large quantities of energy with a relatively low—but not fully exhausted—content of work energy in heat exchangers has potential importance for the optimization of the energy fluxes at the plant. Furthermore, as mentioned, the waste materials from processing may in the future be used to produce biogas or other types of biofuels.

8.4.2 Non-Renewable Energy-Bound Exergy Outputs (NEBEO_INDU)

Most of the non-renewable energy outputs in this sector relate to the consumption of electricity and fossil fuels (mainly diesel). Both are viewed as being fully degraded when used in the production processes, ending up as heat (see the earlier remark about heat exchangers). Thus, an amount equivalent to the sum of the electricity and diesel imports is lost from this sector. For Trolleborg alone the loss is estimated to be 21.1 TJ.

8.4.3 Renewable Matter-Bound Exergy Outputs (RMBEO_INDU)

In principle, all exported goods from this sector may be considered renewable or recyclable. Thus, when using a mass conservation principle and deducting the weight of wastes from the weight of goods we arrive at an export figure for this sector corresponding to 93.7 TJ.

Another amount is assumed to stay within the island boundaries as most of the wastes are used as fertilizers on agricultural land. These wastes represent a work energy value of 7.3 TJ.

8.4.4 Non-Renewable Matter-Bound Exergy Outputs (NMBEO_INDU)

Depending on the non-identified activities that are not included in this analysis, this item may play a varying role. In particular, when the outputs leave the island this strongly affects the possibilities of reuse or recycling on the island. Such a hypothetical possibility is not considered relevant to the situation in 2011, but as the island also wishes to introduce a circular economy this might become relevant in the future.

8.5 Work Energy of ICTS

The work energy stocks, inputs and outputs are shown in Figure 8.1. and are assumed to give at least a rough picture of the situation in 2011. As stated

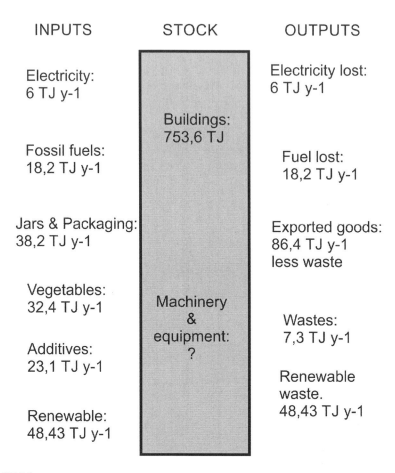

FIGURE 8.1
Diagram showing the work energies in flows and stock of the combined ICTS. The notation has been changed from diagrams of previous chapters as the conversion to work energy has entered directly in the calculation. On the output side, all work energies are considered lost as either exports (goods leaving the island) or the dissipation of energy used in processing.

previously, the roles of some parts of the sector have not been sufficiently clarified.

8.6 Sustainability Indicators of the ICTS

The indicators established in Section 3.8 may now be estimated for the work energies of the ICTS on the island of Samsø 2011.

8.6.1 Stock Indicator of the ICTS

The stock of work energy is estimated from the amount of infrastructure (buildings) to be only 753.6 TJ, as no values could be estimated for the content of machinery and other equipment. This stock is maintained by means of a total input of 117.9 TJ y^{-1}.

Thus, the work energy stock efficiency for the ICTS sector is

$$WESE_{icts} = WE_{stock,\ icts\text{-}sector}/WE_{input,\ icts\text{-}sector} = 753.6/117.9 = 6.4;$$

that is a highly "valued" and necessary sector is maintained at relatively low cost, not considering maintenance and new building activities.

8.6.2 Renewability Indicators of the ICTS

If we consider the fossil-fuel-based energy as the non-renewable input of work energy and the inputs from wind and solar systems together with biomass as the renewable input, we get a renewable to non-renewable index (RNNI) for the sector of

$$RNII_{icts\text{-}sector} = (6 + 38.2 + 32.4 + 23.1)/(18.2) = 99.7/18.2 = 5.5$$

or a renewable efficiency RIEF for the sector of

$$RIEF_{icts\text{-}sector} = 99.7/117.9 = 0.85.$$

The total work energy input from renewable energy sources has been estimated to 99.7 TJ y^{-1}, meaning that slightly more than 85% of the yearly input needed to run the industrial sector is supplied by renewable work energy sources.

8.6.3 O/I Indicator of the ICTS

The output–input efficiency based on work energy of the sector WEOIEF is (assuming that the goods amount to 86.4 TJ and that the wastes have a value of 7.3 TJ, a total of 93.7 TJ):

$$WEOIEF_{icts\text{-}sector} = 93.7/117.9 = 0.79,$$

meaning that 79% of the input work energy is leaving the system.

The energy efficiency calculated as the O/I ratio (times 100) comes to close to 80%. If the wastes—including the potential energies in organic materials—are reused/recycled within the island, this figure will become even higher.

8.7 Sub-Conclusions regarding Industry and More

The attempt made here can only be considered as laying out a strategy for a full analysis of the sector. This is because no real, heavy industry exists on the island. For another society where industries such as steel, cement or similar activities are involved, the method must be elaborated in even more detail. This elaboration will necessarily involve considerations on what may be the right exergy values to use in order to the estimate sustainability of the relevant resources. Are the chemical estimates enough or do we need to involve the cumulative indices—and to what extent should the "embedded" parts consider where the energy inputs necessary to run the processes come from. Such discussions are now taking place in connection to the current project in Jämtland.

9

Work Energy Analysis of Nature

9.1 Introduction to the Work Energy and Eco-Exergy of Nature

From the earlier quoted landscape codes of the Corine Land Cover (CLC) classification system (www.eea.europa.eu/publications/COR0-landcover) the presence of the following areas has been identified from the Geographical Information System (GIS) of the municipality. These landscape data may be considered a slightly more elaborate version of the CLC system. In the calculation of work energy contributed by the systems, a major uncertainty is introduced as neither the codes of the natural systems in the CLC system nor the typology in the local GIS files take into account any variation in the successional states of the systems; that is the landscape is viewed as constant. Thus, at present, it is not possible to take into account the variations in production and biomass with time. This point must be taken as an inherent uncertainty in the method used when determining the work energy contents and flows of nature. The values used represent the average values from each type of ecosystem (mainly from Whittaker, 1975).

The nature areas identified are as follows:

- Forests (aggregation of deciduous, coniferous and mixed stands)
- Moors
- Meadow
- Commons
- Littoral meadows
- Heathlands
- Lakes
- Streams
- Coastal waters

These nature types and their geographical extensions expressed as respective areas have as stated been determined from the polygons in the GIS/MapInfo files available at the municipality.

Additional maps from the municipality account for some marginal areas, such as the boundaries between systems, and include a variety of minor and heterogenous ecosystem types such as the following:

- Hedgerows in the countryside
- User boundaries (no further definition given)
- Hedgerows in cities
- Dikes and dams
- Hillsides

These "systems" are described as lines only (not polygons) and the determination of the spatial extension must in addition rely on a subjective judgement of the respective widths in the landscape. At present, this is the best one can do, but the results must therefore be taken with a considerable amount of uncertainty. Alternatively, a vast number of small ecosystems must be identified and polygons be drawn. Meanwhile, it is highly questionable if the present resolution of maps justifies such efforts with respects to uncertainties.

Most of the work energy in natural systems finds its origins in energy and material of a renewable kind (biomass) due to the free input from photosynthesis. This contribution is unfortunately almost always accounted for as net (primary) production, that is an actual increase in biomass, meaning that the costs in terms of dissipation (respiration of CO_2) occurring in connection to the production has already been deducted.

The stocks of the systems are the standing crop of biomass within the system. As mentioned earlier, no time or space variations are taken into account. Also, standard data from textbooks have to be used since no compilation of data from Danish ecosystems seem available.

The inflows to such a system are limited to primary production as whatever happens in the system of herbivorous, carnivorous or other activities ingestion activities is a result of this primary production. A more precise registration will rely on modern techniques such as remote sensing to estimate the distribution of species, biomasses and production in more details.

The outflows of the system—if not harvested which might be the case of the forests—are reduced to the respirations of the system. In a steady-state system, this must be equal to production. Such a state is only achieved in a special state of the ecosystems known as a climax society (e.g. Odum, 1953).

When considering efficiencies of systems in connections to nature/ecosystems, it only makes senses to calculate the stock indicator as all flow much be considered to belong to renewables either energies or materials. Therefore, only this indicator—an equivalent of the P/B (production per biomass) ratio of ecosystems—has been calculated in for these systems.

9.2 Work Energy of Forests

The forest areas of Samsø sums up to 224 polygons with a total area of 8,812 km²—with some major areas in both the northern and southern parts of the island. Unfortunately, no quantitative maps seem to exist that also includes a distinction among coniferous, deciduous or mixed forest stands, and it has therefore been necessary to treat them as one type only in this report. The information is taken from the maps available at the municipality.

The biomass of forest together with production is estimated as the average of vegetation—32,500 g DW m^{-2} and 1,250 g DW m^{-2} y^{-1}, respectively—in accordance with Whittaker (1975). The same holds for fauna elements, 100 g DW m^{-2}.

9.2.1 The Work Energy and Information of Forests

To convert biomass of vegetation into ecological work energy (eco-exergy) including information a beta value of 325 is adopted. For the fauna element a conversion factor of 200 is used, reflecting that much of the faunal elements will consist of invertebrate species, mainly insects, at a pretty advanced level. For a full accounting of the forest ecosystem, refer to Table 9.1.

9.2.2 Efficiency of the Forest Ecosystems

The forests represent a structure with a WE of 7.75 10^3 TJ and are maintained by an input of 2.06 10^2 TJ. Thus, the S/I efficiency of the system is 37.6. The corresponding value for WE + I is 26.1.

9.3 Work Energy of Moors

The areas of Samsø designated as moors sums up to 92 fragments with a total area of 0.787 km². The moors have been interpreted as wetlands in its widest sense, and no attempt of making a distinction between different types of wetlands has been made. Heathlands are in this context represented by their own type; see Section 9.7.

9.3.1 The Work Energy and Information of Moors

The biomass of moors, together with production, is estimated as the average of vegetation—15,000 g DW m^{-2} and 2000 g DW m^{-2} y^{-1}, respectively, in accordance with Whittaker (1975). The same holds for fauna elements which are set to a biomass of 20 g DW m^{-2}.

TABLE 9.1

Work Energy (WE) and Work Energy Including Information (WE + I) of Forests on Samsø

Estimation of Stocks and Flows of Forests	Value(s)	
Area: 8,811,990 m²	WE	WE + I
STOCK(s)		
Vegetation		
Average biomass density of vegetation in forests: 32,500 g DW m⁻²	5.36 10³ TJ	1,740.5 10³ TJ
(mean, deciduous and evergreen)		
Total biomass of vegetation in forests: 2.86 10¹¹ g DW		
Total C of vegetation in forests: 1.43 10¹¹ g C		
Total work energy of vegetation in forests: 5.36 10¹² kJ = 5.36 10³ TJ		
Total work energy including information of vegetation in forests: 1,740.5 10¹² kJ		
Fauna		
Average biomass density of fauna in forests: 100 g DW m⁻²	1.65 10¹ TJ	3.29 10³ TJ
(mean of fauna deciduous and evergreen)		
Total biomass of fauna in forests: 8.81 10⁸ g DW		
Total C of fauna in forests: 4.40 10⁸ g C		
Total work energy of fauna in forests: 1.65 10¹⁰ kJ = 16.5 TJ		
Total work energy including information of fauna in forests 3.29 10¹² kJ		
Litter		
Average litter density in forests: 14,500 g DW m⁻²	2.39 10³ TJ	2.39 10³ TJ
(mean of litter in deciduous and evergreen)		
Total litter mass in forests: 1.28 10¹¹ g DW		
Total C of litter in forests: 0.64 10¹¹ g C		
Total WE/WE + I in litter of forests: 2.39 10¹² kJ		
Total WE and WE + I in forests	**7.75 10³ TJ**	**1746.2 10⁶ TJ**
IMPORT		
Net production density forests: 1,250 g DW m⁻² y⁻¹	206.0 TJ y⁻¹	66,943.6 TJ y⁻¹
Average production density in C: 625 g C m⁻² y⁻¹		
Total net productivity in forests: 1.10 10¹⁰ g DW		
Total production of C in forests: 0.55 10¹⁰ g C		
Total work energy imported to forests: 2.06 10¹¹ kJ y⁻¹		
Total work energy including information imported to forests: 6,69 10¹³ kJ y⁻¹		

To convert biomass of vegetation into eco-exergy, a beta value of 175 is adopted. For the fauna element, a conversion factor of 200 is used, reflecting that much of the faunal elements will consist of invertebrate species, mainly insects, at a pretty advanced level. For a full accounting of the moors, refer to Table 9.2.

9.3.2 Efficiency of the System

The moors are represented by a structure with and work energy of 294.6 TJ and are maintained by an input of 29.4 TJ y⁻¹. Thus, the S/I efficiency of the system is 10.0, whereas the corresponding efficiency based on the work energy including information is 7.5.

TABLE 9.2

Work Energy (WE) and Work Energy Including Information (WE + I) of Moors on Samsø

Estimation of Stocks and Flows of Moors	Value(s)	
Area: 787,010 m^2	WE	WE + I
STOCKS		
Vegetation		
Average biomass density of vegetation in moors: 15,000 DW g m^{-2}	2.21 10^2 TJ	3.86 10^4 TJ
Total biomass of vegetation in moors: 1.18 10^{10} g DW		
Total C of vegetation in moors: 0.59 10^{10} g C		
Total work energy of vegetation in moors: 2.21 10^{11} kJ = 2.21 10^2 TJ		
Total eco-exergy of vegetation in moors: 3.86 10^{13} kJ		
Fauna		
Average biomass density of fauna in moors: 20 g DW m^{-2}	0.3 TJ	58.9 TJ
Total biomass of fauna in moors: 1.57 10^7 g DW		
Total C of fauna in moors: 0.79 10^7 g C		
Total work energy of fauna in moors: 2.94 10^8 kJ		
Total eco-exergy of fauna in moors: 5.88 10^{10} kJ		
Litter		
Average litter density in moors: 5,000 g DW m^{-2}	73.7 TJ	73.7 TJ
Total litter mass in moors: 3.94 10^9 g DW		
Total C in litter in moors: 1.97 10^9 g C		
Total work energy/eco-exergy of litter in moors: 7.37 10^{10} kJ		
Total (Work Energy) and Eco-Exergy in Moors:	**294.6 TJ**	**3.88 10^4 TJ**
IMPORT		
Net production density in moors: 2,000 g DW m^{-2} y^{-1}	29.4 TJ y^{-1}	5.15 10^3 TJ y^{-1}
Average production density in C in moors: 1,000 g C m^{-2} y^{-1}		
Total net productivity in moors: 1.57 10^9 g DW		
Total production in C in moors: 0.79 10^9 g C		
Total work energy imported to moors: 2.94 10^{10} kJ y^{-1}		
Total eco-exergy imported to moors: 5.15 10^{12} kJ y^{-1}		

9.4 Work Energy of Meadows

The areas categorized as meadow on Samsø sums up to 42 fragments/polygons with a total area of 0.676 km^2. Meadows and littoral meadows have been described as two separate types. They have been treated in quite a similar manner as the commons as they are all grass-dominated lands.

9.4.1 The Work Energy and Information of Moors

The biomass of meadows together with production is estimated as the average of vegetation—1,600 g DW m^{-2} and 600 g DW m^{-2} y^{-1}, respectively, and again in accordance with Whittaker (1975). The same holds for fauna elements, 60 g DW m^{-2}.

To convert biomass of vegetation into eco-exergy, a beta value of 275 is adopted. For the fauna element, a conversion factor of 200 is used, reflecting that

TABLE 9.3

Work Energy (WE) and Work Energy Including Information (WE + I) of Meadows on Samsø

Estimation of Stocks and Flows of the Meadows	Value(s)	
Area: 676,064 m²	WE	WE + I
STOCKS		
Vegetation		
Average biomass density of vegetation in meadows: 1,600 g DW m⁻²	20.2 TJ	5.56 10³ TJ
Total biomass of vegetation in meadows: 1.08 10⁹ g DW		
Total C of vegetation in meadows: 0.54 10⁹ g C		
Total work energy of vegetation in meadows: 2.02 10¹⁰ kJ		
Total eco-exergy of meadows: 5.56 10¹² kJ		
Fauna		
Average biomass density: 60 g DW m⁻²	0.8 TJ	1.52 10² TJ
Total biomass of fauna: 4.06 10⁷ g DW		
Total C in fauna: 2.03 10⁷ g C		
Total work energy of fauna in meadows: 7.59 10⁸ kJ = 0.76 TJ		
Total eco-exergy of fauna in meadows: 1.52 10¹¹ kJ		
Litter		
Average litter density: 3,600 g DW m⁻²	45.5	45.5 TJ
Total litter mass: 2.43 10⁹ g DW		
Total C in litter: 1.22 10⁹ g C		
Total work energy of litter in meadows: 4.55 10¹⁰ kJ		
Total WE and WE + I in Meadows:	**66.5 TJ**	**5.76 10³ TJ**
IMPORT		
Net production density in meadows: 600 g DW m⁻² y⁻¹	7.6 TJ y⁻¹	2.09 10³ TJ y⁻¹
Average production density in C of meadows: 300 g C m⁻² y⁻¹		
Total net productivity in meadows: 4.06 10⁸ g DW		
Total production in C in meadows: 2.03 10⁸ g C		
Total work energy imported to meadows: 7.59 10⁹ kJ y⁻¹		
Total eco-exergy imported to meadows: 2.09 10¹² kJ y⁻¹		

much of the faunal elements will consist of invertebrate species, mainly insects, at a pretty advanced level. For a full accounting of the moors, refer to Table 9.3.

9.4.2 Efficiency of the Meadow Ecosystems

The moors are represented by a structure with and work energy of 66.5 TJ and are maintained by an input of 7.6 TJ y⁻¹. Thus, the S/I efficiency of the system is 8.8, and the corresponding value based on the work energy including information is 2.8.

9.5 Work Energy of Commons

The forest areas of Samsø sums to 70 fragments/polygons with a total area of 6,520 km². The commons are also areas dominated by grasslands, but no clear definition of the distinction has been made.

9.5.1 The Work Energy and Information of Commons

The biomass of commons together with production is estimated as the average of vegetation—1,600 g DW m^{-2} and 600 g DW m^{-2} y^{-1}, respectively, and in accordance with Whittaker (1975). The same holds for fauna elements, 60 g DW m^{-2}.

To convert biomass of vegetation into eco-exergy a beta value of 300 is adopted. For the fauna element, a conversion factor of 200 is used, reflecting that much of the faunal elements will consist of invertebrate species, mainly insects, at a pretty advanced level. For a full accounting of the commons, refer to Table 9.4.

9.5.2 Efficiency of the Commons Ecosystem

The commons are represented by a structure with and work energy of 641.3 TJ and are maintained by an input of 73.2 TJ y^{-1}. Thus, the S/I efficiency of the system is 8.8, and a corresponding efficiency based on the work energy including information of 2.8.

TABLE 9.4

Work Energy (WE) and Work Energy Including Information (WE + I) of Commons on Samsø

Estimation of Stocks and Flows of Commons	Value(s)	
Area: 6,520,003 m^2	WE	WE + I
STOCK		
Vegetation		
Average biomass density of vegetation in commons: 1,600 g DW m^{-2}	195.1 TJ	5.85 10^4 TJ
Total biomass of vegetation in commons: 1.04 10^{10} g DW		
Total C of vegetation in commons: 0.52 10^{10} g C		
Total work energy of vegetation in commons: 1.951 10^{11} kJ = 195 TJ		
Total eco-exergy of vegetation in commons: 5.85 10^{13} kJ		
Fauna		
Average biomass density of fauna in commons: 60 g DW m^{-2}	7.3 TJ	1.46 10^3 TJ
Total biomass of fauna of commons: 3.91 10^8 g		
Total C in fauna of commons: 1.96 10^8 g C		
Total work energy in fauna of commons: 7.32 10^9 kJ = 7.32 TJ		
Total eco-exergy in fauna of commons: 1.46 10^{12} kJ		
Litter		
Average litter density of commons: 3,600 g DW m^{-2}	438.9 TJ	4.39 10^2 TJ
Total litter mass of commons: 2.35 10^{10} g DW		
Total C in litter of commons: 1.18 10^{10} g C		
Total work energy in litter of commons: 4.39 10^{11} kJ		
Total WE and WE + I in commons:	**641.3 TJ**	**6.04 10^4 TJ**
IMPORT		
Net production density commons: 600 g DW m^{-2} y^{-1}	73.2 TJ y^{-1}	2.19 10^4 TJ y^{-1}
Average production density in C of commons: 300 g C m^{-2} y^{-1}		
Total net productivity in commons: 3.91 10^9 g DW		
Total production in C in commons: 1.96 10^9 g C		
Total work energy imported to commons: 7.32 10^{10} kJ y^{-1}		
Total eco-exergy imported to commons: 2.19 10^{13} kJ y^{-1}		

9.6 Work Energy of Littoral Meadows

The littoral meadow areas of Samsø sums to 64 fragments with a total area of 1,617 km^2. Again, this type has been treated as the two previous ecosystems.

9.6.1 The Work Energy and Information of Littoral Meadows

The biomass of littoral meadows together with their production is estimated as the average of vegetation—1,600 g DW m^{-2} and 600 g DW m^{-2} y^{-1}, respectively, and in accordance with Whittaker (1975). The same holds for fauna elements.

To convert biomass of vegetation in the littoral meadows into eco-exergy, a beta value of 275 is adopted. For the fauna element a conversion factor of 200 is used, reflecting that much of the faunal elements will consist of invertebrate species, mainly insects, at a pretty advanced level. For a full accounting of the littoral meadows, refer to Table 9.5.

TABLE 9.5

Work Energy (WE) and Work Energy Including Information (WE + I) of Littoral Meadows on Samsø

Estimation of Stocks and Flows of Littoral Meadows	Value(s)	
Area: 1,617,040 m^2	WE	WE + I
STOCKS		
<u>Vegetation</u>		
Average vegetation biomass density: 1,600 g DW m^{-2}	48.4 TJ	1.33 10^4 TJ
Total biomass of vegetation: 2.59 10^9 g DW		
Total C in vegetation: 1.29 10^9 g C		
Total work energy of vegetation in littoral meadows: 4.84 10^{10} kJ		
Total eco-exergy of vegetation in littoral meadows: 1.33 10^{13} kJ		
<u>Fauna</u>		
Average biomass density: 60 g DW m^{-2}	1.8 TJ	3.63 10^2 TJ
Total biomass of fauna: 9.70 10^7 g DW		
Total C in fauna: 4.85 10^7 g C		
Total work energy of fauna in littoral meadows: 1.81 10^9 kJ		
Total eco-exergy of fauna in littoral meadows 3.63 10^{11} kJ		
<u>Litter</u>		
Average litter density: 3,600 g m^{-2}	1.09 10^2 TJ	1.09 10^2 TJ
Total litter mass: 5.82 10^9 g		
Total C in litter: 2.91 10^9 g C		
Total WE and WE + I of litter in littoral meadows: 1.09 10^{11} kJ		
Total WE and WE + I in littoral meadows:	**159.1 TJ**	**13.8 10^3 TJ**
IMPORT		
Net production density forests: 600 g DW m^{-2} y^{-1}	18.1 TJ y^{-1}	4,989 TJ y^{-1}
Average production density in C: 300 g C m^{-2} y^{-1}		
Total net productivity in forests: 9.70 10^8 g DW		
Total production in C in forests: 4.85 10^8 g C		
Total work energy imported to forests: 1.81 10^{10} kJ y^{-1}		
Total eco-exergy imported to forests: 4.99 10^{12} kJ y^{-1}		

9.6.2 Efficiency of the System

The moors are represented by a structure with a WE of 159.1 TJ and are maintained by an input of 18.1 TJ. Thus, the S/I efficiency of the system is 8.8, and again, a corresponding value based on the work energy including information of 2.8.

9.7 Work Energy of Heathlands

The areas of heathlands on Samsø sums to three fragments with a total area of 2,068 km^2 with some major areas in both the northern and southern parts of the island. Unfortunately, no clear definition of heathlands has been found that justifies the classification.

9.7.1 The Work Energy and Information of Heathlands

The biomass of heathlands together with production is estimated as the average of vegetation—15,000 g DW m^{-2} and 700 g DW m^{-2} y^{-1}, respectively, in accordance with Whittaker (1975). The same holds for fauna elements, 40 g DW m^{-2}.

To convert biomass of vegetation into eco-exergy, a beta value of 300 is adopted. For the fauna element a conversion factor of 200 is used, reflecting that much of the faunal elements will consist of invertebrate species, mainly insects, at a pretty advanced level. For a full accounting of the heathlands, please refer to Table 9.6.

9.7.2 Efficiency of the Heathland Ecosystem

The heathlands are represented by a structure with a work energy of 778 TJ and are maintained by an input of 21.7 TJ y^{-1}, meaning an S/I efficiency of the system is 35.9.

9.8 Work Energy of Lakes and Ponds

The areas of Samsø that has been termed as "lakes" sums to 365 to 478 fragments depending on the data source used and have an estimated total area of 0.581 km^2. It is likely that these systems, in general, also represents smaller ponds, pools and even areas that are only covered by waters at certain times of the year; hence, the difference in numbers of areas identified within this group is simply due to seasonal variations and respective time of observation.

TABLE 9.6

Work Energy (WE) and Work Energy Including Information (WE + I) of Heathlands on Samsø

Estimation of Stocks and Flows of Heathlands	Values(s)	
Area: 2,067,890 m²	WE	WE + I
STOCK		
Vegetation		
Average biomass density vegetation of heathlands 15,000 g DW m⁻²	5.80 10² TJ	1.74 10⁵ TJ
Total biomass of vegetation of heathlands: 3.10 10¹⁰ g DW		
Total C in vegetation: 1.55 10¹⁰ g C		
Total work energy in vegetation of heathlands: 5.80 10¹¹ kJ		
Total eco-exergy in vegetation of heathlands: 1.74 10¹⁴ kJ		
Fauna		
Average biomass density of fauna of heathlands: 40 g DW m⁻²	1.55 TJ	3.09 10² TJ
Total biomass of fauna of heathlands: 8.27 10⁷ g DW		
Total C in fauna of heathlands: 4.14 g 10⁷ C		
Total work energy in fauna of heathlands: 1.55 10⁹ kJ		
Total eco-exergy in fauna of heathlands: 3.09 10¹¹ kJ		
Litter		
Average litter density of heathlands: 5,100 g m⁻²	1,97 10² TJ	1,97 10² TJ
Total litter mass of heathlands: 1.05 10¹⁰ g		
Total C in litter of heathlands: 0.53 10¹⁰ g C		
Total work energy in litter of heathlands: 1.97 10¹¹ kJ		
Total (work energy) and eco-exergy in heathlands:	**7.78 10² TJ**	**1.75 10⁵ TJ**
IMPORT		
Net production density heathlands: 700 g DW m⁻² y⁻¹	21.7 TJ y⁻¹	8.12 10³ TJ y⁻¹
Average production density in C: 350 g C m⁻² y⁻¹		
Total net productivity in heathlands: 1.45 10⁹ g DW		
Total production in C in heathlands: 0.72 10⁹ g C		
Total work energy imported to heathlands: 2.71 10¹⁰ kJ y⁻¹ = 27.1 TJ y⁻¹		
Total eco-exergy imported to heathlands: 8.12 10¹² kJ y⁻¹ = 8.12 10³ TJ y⁻¹		

9.8.1 The Work Energy and Information of Lakes and Ponds

The biomass of primary producers in lakes/ponds together with production is estimated as the average for such systems—200 g DW m⁻² and 500 g DW m⁻² y⁻¹ in accordance with Whittaker (1975). The value for fauna elements is set at a low level—10 g DW m⁻² considering the assumed temporary character of many of these systems.

To convert biomass of vegetation into eco-exergy, a beta value of 325 is adopted. For the fauna element, a conversion factor of 200 is used, reflecting that much of the faunal elements will consist of invertebrate species, mainly insects, at a pretty advanced level. For a full accounting of lakes and ponds please refer to Table 9.7.

TABLE 9.7

Work Energy (WE) and Work Energy Including Information (WE + I) of Lakes and Ponds on Samsø

Estimation of Stocks and Flows of Lakes and Ponds	Value(s)	
Total area of small ponds and lakes (#365): 580816 m^2	WE	WE + I
STOCK(s)		
<u>Vegetation</u>		
Average biomass density of vegetation in lakes: 200 g DW m^{-2}	2.2 TJ	43.4 TJ
Total biomass of vegetation in lakes: 1.16 10^8 g DW		
Total C of vegetation in lakes: 0.58 10^8 g C		
Total work energy of vegetation in lakes: 2.17 10^9 kJ		
Total eco-exergy of vegetation in lakes: 4.34 10^{10} kJ		
<u>Fauna</u>		
Average biomass density in lakes: 10 g DW m^{-2}	0.1 TJ	16.3 TJ
Total biomass of fauna in lakes: 5.81 10^6 g DW		
Total C of fauna in lakes: 2.90 10^6 g C		
Total work energy of fauna in lakes: 1.09 10^8 kJ		
Total eco-exergy of fauna in lakes 1.63 10^{10} kJ		
<u>Litter</u>		
Average litter density: 5,000 g DW m^{-2}	54.3 TJ	54.3 TJ
Total litter mass: 2.90 10^9 g DW		
Total C in litter: 1.45 10^9 g C		
Total work energy of litter in lakes: 5.43 10^{10} kJ		
Total (Work Energy) and Eco-Exergy in Lakes:	**56.6 TJ**	**114.0 TJ**
IMPORT		
Net production density lakes: 500 g DW m^{-2} y^{-1}	5.43 TJ y^{-1}	109 TJ y^{-1}
Average production density in C: 250 g C m^{-2} y^{-1}		
Total net productivity in lakes: 2.90 10^8 g DW		
Total production in C in lakes: 1.45 10^8 g C		
Total work energy imported to lakes: 5.43 10^9 kJ y^{-1}		
Total eco-exergy imported to lakes: 1.09 10^{11} kJ y^{-1}		

9.8.2 Efficiency of the Lake and Pond Ecosystems

The moors are represented by a structure with a work energy of 56.6 TJ and are maintained by an input of 5.43 TJ y^{-1}. Thus, the S/I efficiency of the system is 10.4. Based on the work energy including information, this ratio is 1.0.

9.9 Work Energy of Streams

The "areas" considered to belong to the streams of Samsø sums up to a number of 193 linear fragments. Assuming that these streams are all small and have an average width of 1 m, they can be estimated to occupy a total area of 0.186 km^2.

9.9.1 The Work Energy and Information of Streams

The biomass of streams together with production is estimated as the average of vegetation—1,000 g DW m^{-2} and a low productivity of 100 g DW m^{-2} y^{-1}, respectively, in accordance with Whittaker (1975). The same holds for fauna elements, 10 g DW m^{-2}.

To convert biomass of vegetation into eco-exergy, a beta value of 20 is adopted. For the fauna element a conversion factor of 175 is used, reflecting that much of the faunal elements and the composition of fauna will be in a less advanced state as compared to the more mature terrestrial systems due to the unstable environment. For a full accounting of streams, refer to Table 9.8.

TABLE 9.8

Work Energy (WE) and Work Energy Including Information (WE + I) of Streams on Samsø

Estimation of Stocks and Flow of Coastal Area and Sea	Value(s)	
Total length of small streams (#193): 186,073 km	WE	WE + I
Area: 186,073 m^2 (assuming an average width of 1 m)		
STOCK		
Vegetation		
Average biomass density of vegetation in streams: 1000 g DW m^{-2}	3.5 TJ	69.6 TJ
Total biomass of vegetation in streams: 1.86 10^8 g DW		
Total C of vegetation in streams: 0.93 10^8 g C		
Total work energy of vegetation in streams: 3.48 10^9 kJ		
Total eco-exergy of vegetation in streams: 6.96 10^{10} kJ		
Fauna		
Average biomass density: 10 g DW m^{-2}	0.0 TJ	6.1 TJ
Total biomass of fauna in streams: 1.86 10^6 g DW		
Total C in fauna: 0.93 10^6 g C		
Total work energy in fauna of streams: 3.48 10^7 kJ		
Total eco-exergy in fauna of streams: 6.09 10^9 kJ		
Litter		
Average litter density: 1000 g m^{-2}	3.5 TJ	3.5 TJ
Total litter mass in streams: 1.86 10^8 g DW		
Total C in litter in streams: 0.93 g C		
Total work energy in litter of streams: 3.48 10^9 kJ		
Total (Work Energy) and Eco-Exergy in Streams:	**7.0 TJ**	**79.2 TJ**
IMPORT		
Net production density streams: 100 g DW m^{-2} y^{-1}	0.3 TJ y^{-1}	7 TJ y^{-1}
Average production density in C: 50 g C m^{-2} y^{-1}		
Total net productivity in streams: 1.86 10^7 g DW y^{-1}		
Total production in C in streams: 0.93 10^7 g C y^{-1}		
Total work energy imported to streams: 3.48 10^8 kJ y^{-1}		
Total eco-exergy imported to streams: 6.96 10^9 kJ y^{-1}		

9.9.2 Efficiency of the Stream Ecosystems

The streams are represented by a structure with a WE of 7 TJ and are maintained by an input of 0.3 TJ y^{-1}. Thus, the S/I efficiency of the system is 23.3, with a work-energy-including-information-based 11.3.

9.10 Work Energy of Boundary Zones

Beside the polygons identified as nature in the municipal GIS/MapInfo system, a separate set of systems exist that may be characterized as nature as well. They are all described as line elements only and the determination of their spatial extension therefore depends on a very subjective judgement about, for example, the width of the system.

As no description of the systems nor a proper definition of their characteristics has been fine, it has been decided to aggregate the data and treat them as one type of system with the same properties as meadows described previously (see Section 9.4).

Meanwhile, this type of systems is within recent ecological theory view as particularly important systems. By being part of the boundaries of systems, they become areas with high biodiversity and high activity as a result of the tension between the main type ecosystems, hence the name eco-tones for this type of systems.

9.10.1 The Work Energy and Information of Transitory Ecosystems

The biomass of these systems considered together with production are treated as meadow ecosystems and estimated as the average of vegetation in accordance with Whittaker (1975). The same holds for fauna elements.

To convert biomass of vegetation into eco-exergy, a beta value of 275 is adopted. For the fauna element, a conversion factor of 200 is used, reflecting that much of the faunal elements will consist of invertebrate species, mainly insects, at a pretty advanced level. For a full accounting of these "boundary zones," refer to Table 9.9.

9.10.2 Efficiency of the Stream Ecosystems

The transitory ecosystems represent a structure with a work energy of 424.3 TJ and are maintained by an input of 48.4 TJ y^{-1}. Thus, the S/I efficiency of the system is 8.8 and a work-energy-including-information-based 2.8.

TABLE 9.9

Estimation of the Extension of Various Types of Transitory Ecosystems

"Fence"	Line Elements (#)	Length (m)	Width (m)	Area Estimated (m²)
Hedgerows, countryside	6,259	478,097	2	956,194
"User boundaries"	10,973	1,257,024	1	1,257,024
Hedgerows, cities	3,434	93,724	1	93,724
Dikes and dams	883	194,540	10	1,945,402
Hillsides/slopes	26	3,059	20	61,188
Total area				4,313,532

9.11 Work Energy of Coastal Areas and Sea

The coastal areas surrounding Samsø may be reduced to 1 large polygon that includes the whole municipal area. When the island area is deducted from this polygon the remaining area consisting of shallow waters only is estimated to 1,438 km², more than 12 times the area of the island.

9.11.1 The Work Energy and Information of Coastal Areas and Sea

The biomass of forest together with production is estimated as the average of vegetation in accordance with Whittaker (1975). The same holds for fauna elements.

To convert biomass of vegetation into eco-exergy, a beta value of 325 is adopted. For the fauna element, a conversion factor of 200 is used, reflecting that much of the faunal elements will consist of invertebrate species, mainly insects, at a pretty advanced level. For a full accounting of the coastal area, refer to Table 9.10.

9.11.2 Efficiency of the Coastal Area and Sea Ecosystem

The moors are represented by a structure with a work energy of 6.7×10^4 TJ and are maintained by an input of 2.7×10^4 TJ. Thus, the S/I efficiency of the system is 2.4. Based on the work energy including information, this ratio is 0.5.

9.12 Comparison of Nature Types

Compiling the data from the various types of "natural" systems on allows to elaborate a total estimate of work energy and work energy including

TABLE 9.10

Work Energy (WE) and Work Energy Including Information (WE + I) of Boundary Zones on Samsø

Estimation of Stocks and Flow of Coastal Area and Sea	Values	
Area: 4,313,532 m^2	WE	WE + I
STOCKS		
Vegetation		
Average biomass density in vegetation of "boundaries": 1,600 g DW m^{-2}	129.1 TJ	3.55 10^4 TJ
Total biomass of vegetation in "boundaries": 6.90 10^9 g DW		
Total C in vegetation of "boundaries": 3.45 10^9 g C		
Total work energy in vegetation of "boundaries": 1.29 10^{11} kJ		
Total eco-exergy of "boundaries": 3.55 10^{13} kJ		
Fauna		
Average biomass density of "boundaries": 60 g DW m^{-2}	4.8 TJ	9.68 10^2 TJ
Total biomass of fauna of "boundaries": 2.59 10^8 g DW		
Total C of fauna in "boundaries": 1.29 10^8 g C		
Total work energy of fauna in of "boundaries": 4.84 109 kJ		
Total eco-exergy of fauna in of "boundaries": 9.68 1011 kJ		
Litter		
Average litter density: 3,600 g DW m^{-2}	290.4 TJ	290.4 TJ
Total litter mass: 1.55 10^{10} g DW		
Total C in litter: 0.78 10^{10} g C		
Total work energy of litter in "boundaries": 2.90 10^{11} kJ		
Total (Work Energy) and Eco-Exergy in Boundary Zones:	**424.3 TJ**	**3.68 10^4 TJ**
IMPORT		
Net production density in "boundaries": 600 g DW m^{-2} y^{-1}	48.4 TJ y^{-1}	1.33 10^4 TJ y^{-1}
Average production density of C in "boundaries": 300 g C m^{-2} y^{-1}		
Total net productivity in "boundaries": 2.59 10^9 g DW		
Total production in C in "boundaries": 1.29 10^9 g C		
Total work energy imported to "boundaries": 4.84 10^{10} kJ y^{-1}		
Total eco-exergy imported to "boundaries": 1.33 10^{13} kJ y^{-1}		

information of the ecosystems on the island. For each ecosystem type, the two types of work energies have been divided as either belonging to plants, animals or dead organic matter in the system.

The total work energy of all ecosystems may be found in Table 9.11.

A similar accounting can be found in terms of the work energy including information can be found in Table 9.12.

As seen from a comparison of the two tables (Tables 9.11 and 9.12), the work energy when including information becomes overwhelming and far beyond the order of magnitude of other activities calculated in this book. This is partly because the information included in structures of our society is not included, and to the authors' knowledge, no clearly defined method to do this is found in current literature. Thus, it has been decided for the time being to omit these values (WE + I), although they express an attempt of quantifying what we normally refer to as biodiversity or ecosystem services.

TABLE 9.11

Work Energy (WE) and Work Energy Including Information (WE + I) of Coastal Areas and Sea around Samsø

Estimation of Stocks and Flow of Coastal Area and Sea	Value(s)	
Area: 1,438,000,000 m^2	WE	WE + I
STOCK		
Vegetation		
Average biomass density of vegetation in coastal areas: 500 g m^{-2}	1.34 10^4 TJ	1.34 10^6 TJ
Total biomass of vegetation in coastal areas: 7.19 10^{11} g DW		
Total C in vegetation in coastal areas: 3.60 10^{11} g C		
Total work energy of vegetation in coastal areas: 1.34 10^{13} kJ		
Total eco-exergy of vegetation in coastal areas: 1.34 10^{15} kJ		
Fauna		
Average biomass density of fauna in coastal areas: 10 g DW m^{-2}	2.69 10^2 TJ	4.03 10^4 TJ
Total biomass of fauna in coastal areas: 1.44 10^{10} g DW		
Total C in fauna in coastal areas: 0.72 10^{10} g C		
Total work energy in fauna in coastal areas: 2.69 10^{11} kJ		
Total eco-exergy in fauna in coastal areas: 4.03 10^{13} kJ		
Litter		
Average litter density in coastal areas: 2,000 g m^{-2}	5.38 10^4 TJ	5.38 10^4 TJ
Total litter mass in coastal areas: 2.88 10^{12} g DW		
Total C in litter in coastal areas: 1.44 10^{12} g C		
Total work energy in litter in coastal areas: 5.38 10^{13} kJ		
Total (Work Energy) and Eco-Exergy in Coastal Area and Sea:	**6.7 10^4 TJ**	**1.44 10^6 TJ**
IMPORT		
Net production density in coastal areas: 1,000 g DW m^{-2} y^{-1}	2.69 10^4 TJ y^{-1}	2.69 10^6 TJ y^{-1}
Average production density in C in coastal areas: 500 g C m^{-2} y^{-1}		
Total net productivity in coastal areas: 1.44 10^{12} g DW		
Total production in C in coastal areas: 0.72 10^{12} g C		
Total work energy imported to coastal areas: 2.69 10^{13} kJ y^{-1}		
Total eco-exergy imported to coastal areas: 2.69 10^{15} kJ y^{-1}		

9.13 Comparing the Stock Indicators

The stock indicators were calculated for each type of ecosystem in the previous sections and are here compiled in a table for convenience (Table 9.13)

As seen from the table some systems like forests, heathlands, lakes and boundary zones are maintaining a rather high biomass at relatively low costs. This seems quite logical although there are two reasons for this. The two first may be considered as close to climax conditions, whereas the latter

TABLE 9.12

The Work Energies (WE) of Various Ecosystem Types on Samsø (i.e. before conversion to eco-exergy)

Components	WE in Plants	WE in Animals	WE in Litter	Total WE in Type
Ecosystem Type	(TJ)	(TJ)	(TJ)	(TJ)
Forests	6,329.2	16.5	2,389.4	8,735.1
Moors	140.5	0.3	73.6	214.4
Meadows	20.2	0.8	45.5	66.5
Commons	212.8	7.3	438.9	659.1
Littoral meadows	48.4	1.8	108.9	159.1
Heathlands	632.8	1.5	197.2	831.5
Lakes	0.2	0.1	54.3	54.5
Streams	0.8	0	3.5	4.3
Hedges, etc.	129.1	4.8	290.4	424.3
Coastal waters, etc.	4,889.2	201.7	53,781.2	58,872.1
Totals	12,403.1	234.8	57,382.8	70,020.7

TABLE 9.13

The Eco-Exergies (EE) of Various Ecosystem Types on Samsø (i.e. after conversion from work energy)

Components	EE in Plants	EE in Animals	EE in Litter	Total EE in Type
Ecosystem Type	(TJ)	(TJ)	(TJ)	(TJ)
Forests	1,740,533.3	3,295.7	2,389.4	1,746,218.3
Moors	38,632.4	58.9	73.6	38,764.8
Meadows	5,562.7	151.7	45.5	5,759.9
Commons	58,523.6	1,463.1	438.9	60,425.6
Littoral meadows	13,305	362.9	108.9	13,776.7
Heathlands	174,012.9	309.4	197.2	174,519.5
Lakes	43.4	16.3	54.3	114
Streams	208.8	6.1	3.5	218.3
Hedges, etc.	35,491.7	968	290.4	36,750.1
Coastal waters, etc.	1,344,530	40,335.9	53,781.2	1,438,647.1
Totals	3,410,843.8	46,967.8	57,382.8	3,515,194.4

two can be considered as eco-tones, with intensive gradients able to support at a relatively high biomass. The rest of the ecosystems seem intermediate with indicators around 10. The lowest values are for the coastal areas (2.5–3), which is probably an artefact caused by the conversion of the method to deal a three-dimensional system where organisms and hence biomasses tend to become dispersed resulting a low biomass density.

9.14 Sub-Conclusions regarding Nature

While it has been demonstrated that the route taken to estimate the total value of natural systems in terms of work energy is feasible—a number of compromises had to be made. This is concerned, in particular, with the quality of data around biomasses and production in the systems where old textbooks in the area had to be used. It is probable that very intensive literature studies would reveal more accurate data, maybe even for Danish ecosystems, but the time demand to carry out such a task would be far beyond the limits of the project. In the future, it is likely that for estimation of the autotrophic components (plants: trees, bushes, herbs), new techniques will be available so it is possible by remote sensing to determine species composition, biomass and production simultaneously for the systems and have unique data in time and space for each ecosystem type. Meanwhile, for the time being, economy is limiting for the full-scale implementation of these techniques.

10

Solid Waste—Estimating Amounts and Potentials

10.1 Introduction to Waste Potentials

Although solid waste was intended to be addressed only indirectly, in the project proposal the amount and quality of data from the municipal waste sorting facility made it possible to do a minor but relatively precise overview of the situation in the area. Such an inventory could serve as a basis to a work energy estimation, and an attempt to do so has therefore been carried out in connection to the project. The of the amounts of various types of waste have been accounted for as well as the current practised ways of disposal. Altogether, the data available serve to give a rough estimate of the work energies lost in this manner as well as present possibilities to the future realization of a planned circular economy, that is the potentials for reuse and recycling of valuable resources.

During recent years, the solid waste from the island of Samsø has been collected and partly also sorted at the solid waste treatment facility known as "Harpesdal". At the centre today, only a real minor part is deposited, and the larger amount of solid waste is exported to the mainland (Jutland) for further handling and treatment in the municipalities of Århus and Skanderborg. Thus, upon export, waste may end up being recycled or incinerated outside the island system under consideration.

Collection of material readily available to recycling initiatives, such as glass and newspapers, is practised, but no data for the amounts collected have been identified during this project.

The total refuses handled at the treatment facility for solid waste, Harpesdal, are already on delivery sorted and separated in 10 fractions:

- Garbage (from households)
- Refuse, large (from households mostly)
- Garden garbage
- Industry, trade and commerce

- Sludge
- Building materials
- Bitumen
- Plastics, a major part from agriculture
- Concrete and stones
- Soils

The preceding fractions reflect the system as administered in the years around 2011.

Most of these fractions may be considered as reasonably homogenous, like, for instance, the wastes such as building materials and partly also the wastes from the industry trade and commerce sector.

Meanwhile, this is not valid to the garbage fraction (collected by the municipality on a weekly basis) and large refuses (collected at longer intervals or delivered directly to the plant) which both may be quite diverse in character. The relatively homogeneous fractions may be converted to work energy by using the values for various dominant materials identified and used to calculate the value of infrastructure, whereas the heterogeneous fraction is more troublesome and will need to receive closer attention, as seen in the following.

The potential role of solid waste has been accentuated in the plans for future handling and reuse of materials on the island being currently under elaboration. Waste must be viewed as a resource; some of it may re-enter in other cycling systems on the island, for instance in connection with the biogas production under consideration. Other fractions of more advanced material, such as electronics, may at best be integrated and recycling in activities at larger regional or national scale.

10.2 Municipal Waste and Garbage

The everyday refuses received primarily from private households at Harpesdal sums to 1,760,800 kg during 2011, which corresponds to approximately 350 kg cap^{-1} for a population including tourists, or 475 kg cap^{-1} excluding tourists.

In order to determine the work energy content of the refuses, it is necessary to know the fractional composition of the garbage at least in terms of its major fractions. Unfortunately, no values for the composition of Danish garbage have been identified so far. Instead, values for deposition material in Sweden given in Finnveden et al., 2005 have been used, as the composition is assumed to be much the same in the two countries (see Table 10.1).

TABLE 10.1

The Fractional Composition in Garbage from Households, according to Finnveden et al. (2005), together with an Estimate of the Dry Weight, Exergy Density and Exergy of Fraction and Eventually the Exergy Content per Kg Garbage (values are cumulated exergies in materials)

Fraction	Value	Dry Matter Fraction	WE Cum. (MJ kg⁻¹)	WE in Fraction (MJ kg⁻¹)
Food waste	0,1	0,5	19	1,4
Newsprint	0,4	1	60	25
Corrugated board	0,2	1	60	12
Mixed cardboard	0,1	1	60	6,9
PE	0	1	92	8,3
PP	0	1	92	1,1
PS	0	1	92	0,9
PET	0	1	92	0,4
PVC	0	1	92	0,3
Total				56

Source: From Finnveden et al. (2005).

Note: WE: work energy; PE: polyethylene; PP: polypropylene; PS: polystyrene; PET: polyethylene terephthalate; PVC: polyvinyl chloride.

The values for the work energy densities used in the calculations here are the same that was used for the calculation of work energy in buildings (see Chapter 4). Because of the previously mentioned lack of data for the fractions of glass (bottles and jars) and paper (newspapers and journals) are not included in the calculations.

According to measures made at the collection place paper/cardboard fraction makes up almost 47% of the garbage, which indicates that they may be derived from a system with only little or non-efficient paper/cardboard collection. This also means that this fraction has now in this manner been indirectly included in the calculations as well. Such a calculation might differ much in a system where a separate collection system has been established, a situation normally found in larger cities of Europe.

Given the value of garbage amount earlier—1,760,800 kg—this quantity corresponds to a work energy density of 56.26 MJ kg⁻¹ and a total of 99.1 TJ.

10.3 Building Waste and Garbage

From the values given in the calculation of infrastructure, it is also possible to recalculate a work energy for the wastes of building materials returned to and received at Harpesdal.

TABLE 10.2

The Fractions of Buildings Together with Work Energy Density and the Calculated Work Energy Density in Fractions of Building Waste Received by Harpesdal

Compositional Item	Fraction (#)	WE Density (MJ kg⁻¹)	WE Fraction (MJ kg⁻¹)
Stones and sand	0.1229	1.7	0.209
Concrete and mortar	0.7172	1.7	1.219
Tile and clinker	0.1039	0.75	0.078
Metals	0.014	200	2.791
Wood	0.0265	18.7	0.495
Cardboard and linoleum	0.0006	59.9	0.033
Mineral fiber	0.0083	21.1	0.175
Plastics	0.0018	91.9	0.163
Glass	0.0015	21.1	0.032
Bitumen products	0.0019	40	0.076
Paint	0.0015	1.7	0.003
Total			5.3

Meanwhile, it must be assumed that waste building material is representing a work energy content which corresponds to the average of the buildings. From the same data, it is likewise possible to derive an average fractional composition of buildings composed of 11 different fractional items. This is shown in Table 10.2, and 1 kg of building material is now estimated to have a work energy value of 5.3 MJ kg⁻¹.

10.4 Work Energy in Total Wastes

Most materials received as waste is, at least for the larger and more homogenous parts, being transported to the mainland on trucks for further treatment, but the exact destiny of these materials is unknown. Using the values for work energy densities in fraction derived in Sections 3.2 and 3.3, the work energy content of waste fractions exported from the island in 2011 may be estimated.

It is assumed that the large refuses have the same composition as normal garbage and close to that of wastes from the industry, trade and commerce sector.

The respective values, amounts and densities are shown in Table 10.3.

The garbage collected from private households is by far the most dominating part of the work energy (53.1%) followed by large refuse (assumed too to be from private households) (20%) and the wastes from industry, trade and commerce (15.1%) all responsible to 88.2% of the wastes.

TABLE 10.3

The Amount of Wastes in the 10 Different Fractions Handled at Harpesdal in 2011, Together with Their Assumed Content of Dry Matter, the Respective Work Energy (WE) Densities and the Derived WE Content in the Fractions

Type	Amount	Dry Content	WE Density	In Fraction
Fraction	kg	#Fraction	MJ kg^{-1}	TJ
Garbage	1,760,800	1	56.3	99.13
Large refuse	663,520	1	56.3	37.36
Garden garbage	528,780	0.8	18.7	7.91
Industry, trade and commerce	469,840	1	59.9	28.14
Sludge	193,980	0.9	18.7	3.26
Building materials	611,080	1	5.3	3.24
Bitumen	3,080	1	40	0.12
Plastics from agriculture	28,340	1	91.9	2.6
Concrete and stones	110,900	1	1.7	0.19
Soils	2,735,120	1	1.7	4.65
Total				186.6

Note: Dry content of garden garbage is estimated, while the value for sludge is taken from Valderrama et al. (2013).

Giving a total of 186.6 TJ of work energy lost from the island in this manner; that is exports of wastes represents a significant contribution to losses of resources. None or only very little of these resources are produced on the island, so originally, they must be reflected by the same amount being imported. This point only stresses the importance of reuse and recycling to the island.

The value corresponds to 23% of the work energy exported for instance of crops and 20% of the work energy from wind power produced on the island. While stated in this manner the amount may sound as if it would be of minor importance to the overall budget, it still must be considered that all these exports need to be matched with a corresponding import of materials mostly being of a non-renewable kind. What also becomes interesting to such considerations is the turnover time of components. This may become increasingly important to the fractions of waste where we tend to increase turnover, that is to decrease retention time such as is the case with for instance electronic equipment.

In the end, talking about efficiency in connection to waste component of a system is almost by definition impossible, as wastes—at least when lost, that is not recycled or reused, is the very essence is an expression of the inefficiency of the system.

No estimates of composition of wastewater were made as no data were found to be available. Meanwhile, this component may in future turn out to be valuable as a resource in connection to biogas production.

10.5 Sub-Conclusions regarding Wastes

In the case presented here, a set of relatively precise and reliable data for the quantity and quality of wastes could be obtained. This level of quality of data allows for some considerations on the potential values of for instance recycling of building or construction materials in the society with respect of planning for a future with minimum use of new resources. Meanwhile, at the level of carrying out an evaluation of the potentials connected to the possible use of organic fractions in the biorefinery process, the knowledge is far from sufficient. Likewise, more information is needed on the possibilities to connect to wastes from the industrial sector, such as organic wastes from crop handling activities, dairy and slaughtering wastes, as well as data about municipal wastewater, as a potential resource must be obtained to get a full picture of the situation.

11

Work Energy Budgets—
Overview and Discussion

11.1 Introduction to Discussion and Comparative Overview

Although the aim of establishing a rigorous and general methodology which may serve to make a sustainability of a society at almost any scale has been established, the whole process of literally putting the bits and pieces together in a common framework revealed some essential points that must be taken into account when establishing future activities in this new field of sustainability assessment.

This does also mean that the analysis is not yet complete as it has turned out that for some items, an easy way of estimating work energies cannot readily be identified. This is valid for many of the statistics where the only currency used is money. Here intensive studies are needed to identify these items in terms of amounts which unfortunately will vary with time. Also, a question is raised as to where to put the boundaries in terms of Life-Cycle-Assessment state of a given good being produced or used. Should a substance or element be accounted for in terms of its mere content of chemical work energy, or should we include in the evaluation also the work energy costs is has taken to produce that state of the element, for example iron content in the form of an ore (natural value) or iron content after its extraction, melting and other processing (a cumulative, embedded value)? A big part of the environmental burden of a product may turn out to be placed outside the system.

Before returning to a comparative evaluation of the relative importance of the sectors on the island of Samsø, we attempt to make a brief and condensed description of the major obstacles to the construction and implementation of such a new methodology as it has been attempted here.

The outline of the method was relatively clear from the beginning, that is to establish a framework based on work energy which would serve to map how flows of work energy—bound in energy and matter—were distributed through our society. Such a mapping would serve to identify where big investments and consumption occur in our society and thus would assist us in determine where to act and implement measures.

Whereas the previous scope was clear it turned out that the unravelling of the necessary data to carry out the mapping was not always as easy a task as first believed. It seems that in our society, we tend to gather and cumulate data in great detail and with relative ease, but once data has been collected, they are not always available in a consistent form. They are, for instance, very much dependent on from what sources they are derived, and it is sometimes difficult to get them in a suitable form. It is necessary to address such issues in the future and coordinate between the many levels of authorities if this part of the process should be facilitated. Meanwhile, it is also recognized that this is not a problem that is particular for this specific method. Rather the lack of consistency in environmental (cost) accounting poses an obstacle to all initiatives in the direction of carrying out a sustainability evaluation of a society. A first step has been taken by integrating energy and material fluxes by means of a method based on work energy.

11.2 Overall Issues

Of course, it will be nice to have an overall figure to describe the sustainability level of a given society. Such a figure may be observed over time, and if chosen properly, the figure will serve to indicate if conditions have improved and the society is changing its performance in the direction of being more sustainable. The consequences of measures taken can be seen more or less directly and instantaneously.

Meanwhile, one overall figure is hardly enough to direct our efforts, and therefore, more detailed knowledge as in the approach presented here is needed. Hence, the society has been divided into some important sectors present in almost all societies, although the sectors will differ in their respective role and importance between societies. Thus, important sectors must be identified, and following this identification, more intensive and detailed investigations can be initiated like, for instance, in case of societies with heavy industries, fisheries as dominant activity and so on. The division of sectors in this report is believed to be general as it is a condensate of approaches found in current literature and statistics.

At the centre of the development of the method is also placed a view of societal activities as being spatially correlated; that is all activities we as humans undertake is taking up geographical space. The rest of the space of a given area that is not directly involved in human activities we identify as nature. In principle, this partitioning should also be valid to all countries and will serve to reveal within-country variations, such as the high storages in cities as well as the role of ecosystems services in the countryside.

11.3 Issues Related to All Sectors

This section is concerned with methodological problems that penetrate the analyses of all sectors:

1. Determination geographical extension—which deals with the extension of area used for human activities to nature
2. Buildings and infrastructure—and their related costs
3. Materials and chemicals—the flows in and out
4. Information—another set of costs that are rarely estimated or evaluated

11.3.1 Determining Geographical Extensions

The identification of the geographical extension of all activities has been roughly grouped according to the classification system set up by the previously mentioned Corine Land Cover system which is defining types of landscape which altogether belongs to either cultural activities or to natural systems. Taking this system as a first entrance point should make the approach compatible with activities undertaken by most European countries or members of the European Union. Many countries today have established a similar system that it will be possible to use in connection with this method.

Meanwhile, the data in the Corine system are becoming "outdated". The latest available data used for this analysis were established in 1996. Also, the spatial resolution of these landscape data is too rough to carry out a sustainability analysis of smaller societies such as municipalities. Meanwhile, the general development in this field has been so that an increasing number of municipalities have started their own initiatives to get a better overview of their own areas.

The eventual division of landscape and the resolution with which results may be obtained will depend on the availability of high-resolution data. Such data are available for most Danish municipalities and may be obtained by the relevant authorities. Meanwhile, the privatization of previously governmental authorities that have followed the neo-liberal turn of many countries has had the effect that data are often not any longer available for free. The price of sufficient data can amount to a considerable amount and may pose an obstacle to a fully detailed analysis in some areas. This has been found to be the case in both Denmark and Sweden.

The areal estimates make it possible to quantify the part of society taken up by infrastructure—and in principle give an estimate of the space needed for societal activities as well as persistence nature. Thus, it will be possible to give a rough calculation of the immediate stocks of work energy involved in activities and sectors.

11.3.2 Buildings and Infrastructure

One major problem appears as one attempts to convert from the areal extension of building structure into its respective mass and subsequently work energy. This problem is caused by the fact that only a few data seem to exist that (1) first, describe the relative composition of buildings in terms of materials and (2) even has an accounting of this for various types of buildings. The optimal data set identified and used in the investigation was taken from an analysis of buildings which was made with the purpose of estimating the amounts of materials set available by demolition and as thus potentially available as a resource in the implementation of reuse, recycling strategy of the building sector, that is realization of circular economy of this sector.

It turns out, that for most sectors the value of the infrastructure in terms of work energy is quite high. In fact, when using a relatively conservative estimate of a replacement or remodelling rate of 1% per year, one gets a value of such order of magnitude, that it should really lead to us to be much concerned about the potentials in reusing these outputs as resources.

11.3.3 Materials and Chemicals

Next, information on the work energy of materials and chemicals is needed. The transition from elemental composition via more complex molecules to the variety of materials that ends up forming the construction material is non-trivial as work energies tend to be non-additive!

The first issue of getting values for materials is merely a time-consuming process which indicates the necessity of a database containing such data. Most data for this type of analysis were found in reports and papers mainly derived from studies of life cycles of products. The data available from such sources have increased considerably over recent decades.

In general, the cleaner the molecule—the closer to elements—the easier it is to get a value of its work energy content. Here the works of Szargut and colleagues (Szargut et al., 1988; Szargut, 2005) forms an invaluable source of data. In addition, the works also indicate methods to be used for the calculation of the work energy for more complex molecules. Some help may nowadays be found in the "Chemical Exergy Calculator" or "Flow Exergy Calculator" found at the website of ExergoEcology Portal (www.exergoecology.com).

Meanwhile, the general situation leaves the method with a grey zone where more efforts should be taken, and the just mentioned portal represents an initiative supporting this direction. When taking into account that, at present, more than 100,000 compounds have been approved for use in the European Union, it would be necessary, and the next logical step would be to build up a database of work energies of elements, compounds and derived substances. On the way to this, it would be possible to use minor subsets of data in order to determine the best way of calculation.

It is suggested to set up a research project with the purpose of evaluating methods for work energy of compounds and elements comparing existing methods against each other. A database could then be prepared that includes several of the parameters for testing their importance to the results.

11.3.4 With Information?

Much of the infrastructure of our society as described earlier is composed of materials, and we calculate for instance the work energy value of a building as the sum of work energies in its structural components not considering how—more or less—sophisticated they are put together. That is, the total work energy of a structure equals the sum of the work energy of its respective parts of which it is composed. We tend to see society to be a system basically—with the exception of ourselves to be composed of "dead", mostly inorganic components.

The argument may be valid to the extent that we deal with the very basic elements that we consider needed for our existence, such as constructs for houses, clothes and provision of nearby food. But today we or rather our societies seem to "need" much more than just the basic needs given by, for instance, Maslow's pyramid. This trend of a turn towards increasing materialism and dependency of society on items with a long transport route has become much clearer with the so-called globalization entering the scene. In short, our societies have become dependent on transfers of energy, matter and relevant information all over our planet. In particular, many countries in the industrialized world live on the work energy extracted in other, more remote places such as third-world countries.

Furthermore, we consider ourselves to live in an "information society" where we and our adjacent structures such as governments and banks are strongly dependent strongly on the fast and massive exchange of information (internet, media, big data and alike). Strangely enough, no exact data has been found on the value and importance of all these types of "necessary" data.

As opposed to this, nature with its organisms, our surrounding ecosystems are more than dead physico-chemical systems, yet the early calculations of work energy of nature that considered only the chemical composition of the organisms in the ecosystem did not take into account that they were also alive. In other words, living organisms carry out activities which cannot be performed by dead materials. An organism should be considered more than just the value of, for instance, its content of Gibbs free energy. As indicated in Chapter 2 through 4, through studies of ecosystem function it has been found necessary to consider and include also the information content of organisms in measures of the complexity and organization of the ecosystem. As a result, it was suggested that the information content needed for the function of an organism would be reflected in the amount of DNA it possesses. For an overview of this, see Nielsen et al. (2020).

Recent technologies within the science of genetics have made it relatively easy for us to come up with a measure for the information content of organisms at different levels of complexity. Therefore, the measure of work energy with information was also introduced in this treatment.

As seen, the calculation is easy to carry out, but including this aspect in the calculation—that is the work energy with information as opposed to the pure chemical work energy—makes nature to become dominant in terms of work energies—this goes for both nature and agro-ecosystems. For viewing the differences of natural ecosystems, the reader is referred to the tables in Chapter 9.

This issue serves to illustrate the importance of nature. Not only do we need it in terms of quantity, but we also would need to preserve its qualitative aspects—both elements included in what we today refer to as ecosystem services.

For a comparison with societal structures, this may not be fair. Meanwhile, as previously mentioned, so far, no good investigation of information contents or flows in society have been found, and this may be a valuable area for research efforts in the future. Some suggestions may be found in Nielsen (2007), Nielsen and Müller (2009) and Jørgensen (2012)

The coexistence of both society and nature is necessary and realized today. If future investigations in this area are carried out, it will become increasingly important to include considerations on the issues reflected in the debates around whether we should strive for a strong or a weak sustainability. In particular, as long as a proper economic theory that includes and comprehend, such views has not been developed.

11.4 Specific Sector-Related Issues

This section deals with the situation of the particular sectors, both in general and specifically on the island of Samsø in 2011. The general problems of the methodology addressed earlier are not repeated here but refer to the previous discussion.

11.4.1 Energy-Sector Issues

The inventory and budget elaborated by PlanEnergi for the municipality are suitable and adequate for a first evaluation of the work energy budget of the "energy-bound" work energies of the island.

The only weakness is that the budget does not include material work energies. Thus, it does not include considerations on infrastructure and other necessary flows of this kind. Nevertheless, as a starting assumption, hypothetically the structures are not quantitatively important which they actually turn out to be. But at least they are qualitatively important in fulfilling the role of supplying the island with energy. For the future, it is likely that the work energies in material contents and flows to this sector will be of increasing importance.

11.4.2 Public-Sector Issues

The problem of determination of the building infrastructure is inheriting the uncertainties indicated earlier. In some cases, it would be necessary and practical to study this sector in more details, for instance in situations where municipalities are addressing sustainability issues and focusing on their own performance in terms of efficient work energy usage.

Likewise, it would be necessary to obtain work energy measures for particular buildings and activities in order to establish a more goal-oriented action, for example improving the insulation of certain building types. This would require that building and consumption data would be studied carefully and in even more detail. At the moment, it is not easy to tell exactly with what ease such detailed analyses can be performed. In the case of the island of Samsø, an energy advisor has been employed by the municipality for a period dedicated to carrying out such studies.

11.4.3 Private-Sector Issues

Determination of building infrastructure share the previously mentioned problems, and in addition, it is difficult to get data for actual renewal or remodelling rates of buildings or, for that matter, for the construction of new buildings.

That is, it seems difficult to get data for larger changes in the infrastructure, such as construction and condemnation. It is probably possible to get data for new construction sites from the official registries whereas no source for condemnation data has been identified. This area should receive much more attention in the future.

Inside this infrastructure, we tend to place things with great variation in their consumption and replacement rates, from items with long duration (built-in closets, furniture) to "fast-moving consumer goods" (electronics, paper, etc.). For the time being, the consumption of such items has only identified in terms of money used for consumption which is not easily convertible to mass and hence cannot be converted to work energy.

As many of these consumables are used with high rates today and considering that they may contain valuable materials such as copper and other rare and indispensable metals, it may be important in the future to know more about the composition of the items as it will become increasingly important to ensure as perfect a recycling as possible of these materials.

11.4.4 Agriculture-Sector Issues

As an important result, it very early became clear that it was necessary to divide the activities of this sector in yet a number of sub-sectors. Although, at least crop production, livestock production and to a certain extent also forestry activities may be undertaken by the same persons. The activities deviate considerably from each other, have their own coding systems for their products and different approaches and angles on calculating necessary resources and the expected outcomes. Establishing a robust method to this area of activity, it was considered that it would be too complex to establish one single unified system for description of one single farm and that it would be more practical and feasible to split up the activities as illustrated in the report.

11.4.4.1 Crop-Sub-Sector Issues

Farmers involved in Danish agriculture are participating in report systems which make it relatively easy to obtain reliable data with great accuracy once the responsible data source has been identified. This report system in principle contains information where single fields are situated in the landscape, what is grown on the land. They also report about the use of fertilizers, being it natural or artificial, and the use of pesticides. Meanwhile, shortly after the collection of data for this project access to data started to become commercialized as mentioned before; that is data are not freely available any longer.

For the time being, it is not clarified if the data from farmers are reported as part of the same system as the Danish farmers have at least three available systems to enter information about their farming activities. If the systems can be coupled, it will become easier to derive a detailed picture of for instance the sustainability level of different crops or even farms.

It was chosen, during this project to calculate the outcome of crops by using the average values given by farmer advisory organizations and governmental institutions as much as possible. Of course, actual outcomes can also be used to give a more precise picture of a particular year. Meanwhile, this would have forced the project focus even more to the collection of data and reduced time for method development. Meanwhile, for some particular crops of minor quantitative importance only, data from organic farming were available.

The amount of sowing seed input necessary to make a certain outcome was calculated using the equations also given by the advisory functions and literature. The calculation of this amount is dependent on three parameters which are varying between crops, and they are all connected to some degree of uncertainty. Thousand grain weight has been identified for most important species, but for some species, the values for similar varieties had to be used. Likewise, the plant density will depend on considerations of the farmer, soil structure and the success of the sown species in actually turning into a plant which is given by the germination percentage, that is the percentage of seeds successfully converted into plants. The values used here have been calculated, but sometimes a relatively large deviation between calculated values and the values recommended in various materials, home pages and brochures has been noticed. It has not been possible to establish the reason for this.

The consumption of work energies by various soil handling activities have been calculated and described in detail (Dalgaard et al., 2002), and the data from this report have been used to calculate the needed investment for this. For some crops, a relatively detailed description of recommended soil-handling activities has been identified, for example as described in advisory manuals or on home pages. Where such instructions have not been found, it has been assumed that similar crops receive a similar treatment.

For the usage of artificial fertilizers and pesticides, average values have been used. Fertilizers' average use of nitrogen, phosphorus and potassium, as indicated for many crops, has been calculated. Values for use of pesticides have been taken from the yearly report as elaborated by the environmental ministry. A more detailed and specific calculation is not feasible without intensified data search and is dependent on several of the general issues raised earlier.

11.4.4.2 Livestock-Sub-Sector Issues

Even though the livestock production system at a first look seems less complex in terms of numbers of elements, for example the number of types of animals raised, the production itself is a rather complicated affair. Meanwhile, it reaches its complexity in other ways, animals of seemingly the same type may be raised with quite different purposes, for example meat vs. dairy production. This has the consequence that similar animals may receive quite different treatment, grow differently and have different consumption needs and outputs.

The island is relatively simple, but the data used have been taken from 2012 which is, by the veterinarian authorities, stated not to be different from the situation in 2011 (pers. comm.). It was simply at the time when final accounting was carried out not possible to obtain the data set from 2011 any longer. For simplicity, the animals were counted only for stock larger than three entities.

Although not figuring in the data, the presence of a number of horses, sheep, cattle and chicken have been observed on the island. Due to lack of data, they are not included here.

11.4.4.3 Forestry- and Fisheries-Sub-Sector Issues

For other societies and at other scales, forestry and fisheries may be important and require that a more detailed description of the methodology be developed for these sectors. Both activities require a description of how to estimate their respective values in terms of work energy values. An attempt to doing this is at present under development for forestry activities in Jämtland, Sweden (Skytt et al., 2019, 2020).

As the forestry sub-sector at Samsø is considered low in intensity and since activities within fisheries sub-sector have been reduced to private activities for which data are not easily obtained, both sub-sectors have been neglected in this study.

11.4.5 Industry, Trade and Commerce Issues

The situation reported here of the industrial sector—including trade and commerce—is biased by the fact that only one plant—together with adjacent handling and transport activities—has been reported as industry proper. Considering the limited size of the sector structure, this indicates that the method is still not adequate to identify important consumptions even at small scales.

None of the activities on the island qualify for making a mandatory green accounting of the activities. As such accountings have been shown also to be of value to at least medium-sized production units, it could be of value to a small-scale society also to focus on the role of such activities. Recent findings indicate the value of establishing even simple accounting systems to even small and medium-sized enterprises. In the case of implementing the method to systems at a larger scale, it would be obvious to take a starting point at units already making green accountings.

Industrial activities vary a lot but may be systematized according to the consumables that they use. With the work energy values of consumable items, that is raw materials, elements and chemicals, among others, in a database, it would be a relatively simple process to convert a green accounting into its corresponding work energy balances, thereby evaluating different industries with each other.

11.4.6 Nature-Sector Issues

Nature is complex too, and in this context, it can and must be divided into its functional groups—or what scientists mainly refer to as ecosystems. This is important since various types of ecosystems have different biomasses and production, as compared to the 24 principles of E. P. Odum (Odum, 1969) which is not only important to the stocks and imports of work energy—but also very important to the evaluation of the carbon balance of the ecosystem—which is demonstrated in the adjacent carbon model developed for the island (Jørgensen and Nielsen, 2015). Such ecosystems differ not only between each other but also within their life cycle; they may perform with considerable variation—not to mention the yearly variations. The latter variation is not important when a yearly time scale is used for the implementation of the analysis.

Unfortunately, accurate data for typical Danish ecosystems are hard to obtain, and most researchers often refer to standard work in the area that covers ranges of data for ecosystems from all over the world, for example Whittaker (1975). As estimates, personal judgements of the values given in this work were used. The values given are with wide ranges and covering only plants, smaller heterotrophs and litter, all which may be essential to both work energy and carbon relations.

Thus, even though the estimation must be connected to high uncertainties, the work energies of nature are important as storages and importance, and it is likely that the importance is underestimated. It should also be considered that the coastal ecosystem as treated here may be even more important in shaping the total contribution from ecosystems services than indicated, for example support of pollinating insects.

If estimating the value of nature as "work energy including information" (WE + I), its value becomes overwhelming—a point that is only stressing the value of nature. With information, namely the functionality of organisms and nature is included, it is this functionality which is necessary for nature to carry out all the ecosystem services it offers.

11.4.7 Waste-Sector Issues

A considerable amount of work energy leaves the island as wastes every year, roughly estimated to correspond to around 45% of the sustainable energy import to the island. This means that almost half of imported work energies sooner or later end up as waste.

So far, only a little attention has been paid to these in some cases valuable resources leaving the island. Future concerns should invest efforts in keeping as much of the resources on the island as a possible subsidiary to ensure proper recycling elsewhere outside the system.

11.5 Comparison of Sectors

A comparison of the sectors may occur at two levels. The first deals with a comparison of their stocks, inputs and outputs of work energy. This will serve as a first quantitative evaluation and will reveal the investments required to maintain the function of the sectors as well as to give a first intuitive picture of their efficiency. Here it should be noted that irreversible losses of work energies are bound to happen. They cannot be avoided, only reduced. The second level is to compare their overall efficiency ratios as presented in the report.

11.5.1 Work Energies in Stocks

This chapter includes considerations on how to integrate the various proposals on systematic and systemic approaches to the analysis of sustainability in regions and society.

The order of magnitude of nature is comparable to the infrastructure of some of the cultural sectors. Depending on the turnover ratio of the renewable materials on the island, this result could be interpreted as if the nature of Samsø, in principle, could be able to supply renewable structures on the island. Meanwhile, much infrastructure that hitherto has been based on non-renewable materials then needs to be replaced. This immediately raises the question of whether a replacement can be realized at all and then on what time scale.

When evaluating nature in the form of work energy including information, it becomes clear how important it is to maintain this infrastructure together with its information. The information part represents the functionality and biodiversity and, in the end, expresses the value of ecosystem services.

11.5.2 Work Energies in Inputs

This part of the budget serves to reveal places where large inputs are needed and/or created for the system or a specific sub-system to sustain.

The agricultural sector is dominant when it comes to inputs required for the sector. Meanwhile, most of this work energy is delivered from solar radiation and hence is derived from a renewable resource. What eventually will

TABLE 11.1

Values of Work Energies in Stocks of Samsø in 2011 Divided for All Treated Sectors and for the Four Types: (a) Renewable Energy Bound, REBES; (b) Non-Renewable Energy Bound, NEBES; (c) Renewable Matter Bound, RMBES; and (d) Non-Renewable Matter Bound, NMBES (all values in TJ)

Sectors WE Types	Energy	Public	Private	Crop.	Livest.	ICTS	Nature	Totals	Remarks
REBES	–	–	–	–	–	–	–		Energy storage technologies not implemented
NEBES	–	–	–	–	–	–	–		do
RMBES	–	368	2,187	3,388	4,0	7,54	10,196		Mainly buildings and biomass
NMBES	–	8,823	ne	1,568	ne	ne	–		do
Totals	–	9,191	2,187	4,956	40	754	10,196		

Note: RMBES of energy is to a large extent likely to be included in either the public- or private-sector data. For the NMBES, some values could not be estimated from the information given—but probably are included as part of RMBES. For nature, 6,562 TJ accounts for primary producers. ne = not estimated.

TABLE 11.2

Values of Work Energies in Inflows of Samsø in 2011 Divided for All Treated Sectors and for the Four Types: (a) Renewable Energy Bound, REBES; (b) Non-Renewable Energy Bound, NEBES; (c) Renewable Matter Bound, RMBES; and (d) Non-Renewable Matter Bound, NMBES (all values in TJ)

Sectors WE Types	Energy	Public	Private	Crop.	Livest.	ICTS	Nature	Totals
REBEI	442	10	93	3,377	9	6	–	
NEBEI	284	53	78	31	?	18	–	
RMBEI	94	4	31	30	376	32	416	
NMBEI	–	406	13	57	?	61	–	
Totals	820	473	215	3,495	385	117	416	

Note: ? = not estimated mostly due to lack of data.

turn out to be critical in the context of sustainability are the inputs used, which are based on non-renewable resources.

The import of work energy to nature is in an order of magnitude comparable to sustainable energy created on the island.

11.5.3 Work Energies in Outputs

As stated earlier, many of the losses are unavoidable—but large amounts of work energies are lost from the sector system—or, for that matter, any of the sub-system can be of potential interest with the purpose of optimization, minimization or even elimination (by excluding the activity or process). Anyway, the introduction of such initiatives and measures are dependent on a long line of factors, from the decision process, actual choices made and very much on the available options of new technologies at hand.

Table 11.3 reveals two types of loss of potential interest—namely the unavoidable losses, mainly from energy, and some material losses.

The part of the material losses that could be possibly retained on the island and used in recycling activities are of course of particular interest. Major non-reversible losses should be checked for the apparent lack of efficiency of the sectors. A thorough analysis and search for causes and sources will in some cases be necessary.

11.6 Comparing Indicators

The indicators chosen are quite different in character—in fact, only the latter of the two proposed indicators, referred to as efficiencies—have clear developmental trends to follow during monitoring.

TABLE 11.3

Values of Work Energies in Outputs of Samsø in 2011 Divided for All Treated Sectors and for the Four Types: (a) Renewable Energy Bound, REBES; (b) Non-Renewable Energy Bound, NEBES; (c) Renewable Matter Bound, RMBES; and (d) Non-Renewable Matter Bound, NMBES (all values in TJ)

Sectors WE Types	Energy	Public	Private	Crop.	Livest.	ICTS	Nature	Totals	Remarks
REBEO	316	0	0	0	0	0	–		
NEBEO	–	0	0	0	158	0	–		
RMBEO	–	–	22	44	162	7	–		Exported – but partly useable
NMBEO	–	410	0	376	38	86	–		Exported and lost
Work energy Loss Totals		53	217	106	–	24	416!		

Note: A relatively large amount of manure/slurry is emerging from livestock production—which may be diverted to fields or biogas production. The high work-energy loss of nature assumes a steady state of the primary producers—which should probably be considered as productive and input only. ICTS = industry, trade, and commerce sector.

The stock indicator does not necessarily have a clear trend (limit), and its trend will depend on actions taken according to the rule of minimizing stocks and minimizing their respective costs, for example of maintenance. In fact, the actions taken in many societies aiming at replacing an old structure with a new structure believed to be better will often require an investment of non-renewable materials and make the indicator look bad at a first glance.

The renewability indicator—the ratio between renewable to non-renewable resources used for inputs—will probably lose its meaning, as non-renewable resources are getting low and the indicator therefore runs to infinity. In this case, uncertainties in data used for calculation are likely to become dominant and the following should be used.

The renewability efficiency is likely to be operative during this phase of transition, that is when the society is really approaching sustainability by having all the resources driving it are becoming of a renewable kind, that is when the efficiency approaches one (1).

The output–input (O/I) indicator will also be variable and follow certain trends which may be observed together with actions taken.

In fact, the stock/input indicator and the output/input indicators are predicted to show a behaviour which may only be fully understood when taking into account the implemented measures and the political wishes lying behind actions, like for instance the previously mentioned example of investment in changing the infrastructure.

TABLE 11.4

Values of Various Sustainability Indicators Suggested and Applied to the Analysis of the Island of Samsø in 2011

Sectors WE Types	Energy	Public	Private	Crop.	Livest.	ICTS	Nature	Totals
Stock indicator	–	19.5	40.2	1.41	0.1	6.4	–	
Renewability indicator	1.9	0.03	1.6	38.9	ne	5.5	–	
Renewability efficiency	0.65	?	0.65	0.97	1	0.85	–	
O/I indicator	0.39	0	0.11	0.35	0.51	0.79	–	

Note: ICTS = industry, trade and commerce sector; O/I = output/input; ne = not estimated; ? =.

The stock indicator seems to be high when human infrastructures are involved, while the greater the importance of "nature" in the system under consideration, the lower the value.

The renewability indicator shows almost the opposite trend, but the data are too few to determine whether this is also a general trend.

The *renewability efficiency* is close to one for the two agricultural sub-sectors, that is the sectors based on natural resources—whereas the societal sectors seem to follow the energy-sector efficiency.

The output/input efficiency (O/I indicator) is highest for the sectors that either export energy (the energy sector) or the sector which exports work energy bound in materials. The lowest values are found in the public and private household sector.

11.7 Work Energy and Samsø in 2011

When aggregating the data for the island and concentrating on fluxes that pass over the boundaries, we may get an overall picture of the situation. The cumulated values are shown in Figure 11.1.

The work energies imported to the island adds up to a total of 1410 TJ y^{-1}. When splitting into energy or matter the contributions are estimated to be 726 TJ y^{-1} (51%) and 684 TJ y^{-1} (49%) as energy- or matter-bound work energies, respectively. Out of this, 599 TJ y^{-1} (42%) are considered to originate in renewable resources, and 811 TJ y^{-1} (58%) are derived from non-renewables.

The outputs sum to a total of 1,603 TJ y^{-1}. Out of this, 316 TJ y^{-1} (20%) represents the export of the excess energy produced from windmills. The remaining 1,287 TJ y^{-1} (80%) are all material-bound work energy mainly in the exported goods (agricultural products) and solid wastes. From the total,

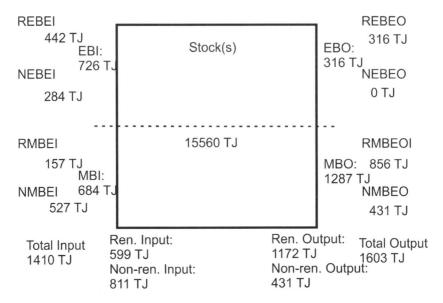

FIGURE 11.1
An aggregated representation of all flows and stock on the island in 2011

1,172 TJ y^{-1} (73%) are considered to be renewables and 431 TJ y^{-1} (27%) to be non-renewables.

The overall balance is when calculated as input less output −193 TJ y^{-1}, meaning that more is exported than invested in the society. Considering that a society should have a work energy balance greater than or equal to 0, the system cannot be evaluated as sustainable.

The indicators for 2011 come out with the following values:

Stock indicator = 15,560/1,410 = 11.0

Renewable/non-renewable ratio = 599/811 = 0.73

Renewability efficiency = 599/1,410 = 0.43

O/I indicator = 1,603/1,410 = 1.14

The stock indicator shows that the infrastructure is maintained by an input corresponding to 11% of its size per year.

The renewability ratio shows that renewable resources have 73% the size of non-renewables—when converted into renewable efficiency, renewable resources are responsible for 43% of the inputs of work energy.

The O/I indicator shows that more work energy is leaving the society than entering it. This is not necessarily a negative. First of all, the electricity exported is contributing to the sustainability of adjacent societies. Also, the input as accounted here are not including the input from solar radiation which is responsible for making the exported material goods possible.

11.8 A Scenario for 2020

A hypothetic scenario for the situation in 2020 may be built on a reorganizing the consumption pattern in accordance with existing possibilities and known technologies. As the islanders have a plan to become independent of fossil fuels, it is assumed that the substitution of non-renewable energies will have the most attention.

11.8.1 Supply-Side Considerations

When looking at the imports it is assumed that the role of renewable energies from wind power, photovoltaics will increase over the years by 5%, or 22 TJ, to a total of 464 TJ y^{-1}.

The non-renewable work energy inputs in 2011, according to the preceding, sums up to 284 TJ y^{-1} and stems from transports: ferries, 161 TJ y^{-1} (diesel), diesel cars, 17 TJ y^{-1}, and other cars, 39 TJ y^{-1} (gasoline); heating, 47 TJ y^{-1}; and industrial activities, 18 TJ y^{-1}. Within the 2013–2020 period, it should be realistic to replace work energy consumption by cars, heating and industry by electric equipment, corresponding to 121 TJ y^{-1}.

The consumption by the new ferry *Koldby-Kås/Kalundborg* will also be replaced by renewable fuels. This may reduce the imports of fossil fuels by an additional 60 TJ y^{-1}, all in all a reduction of 181 TJ y^{-1}. Consequently, the

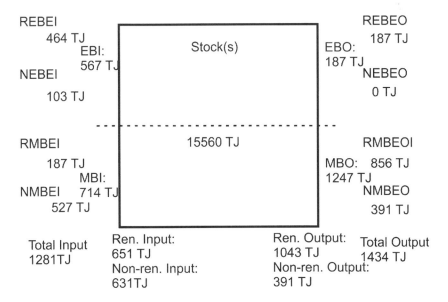

FIGURE 11.2
A possible scenario for the flows and stocks on the island in 2020 after introduction of changes

amount of renewable matter-bound work energy must increase by an estimate of 30 TJ y^{-1}, if assuming a partitioning of energies between biofuels and electricity to be equal.

Everything else is assumed to be equal. Meanwhile, it should be noted that a conversion to organic farming would reduce the required input of the island by at least 57 TJ y^{-1}.

11.8.2 Output-Side Considerations

The exports of work energy are to a large extent composed of contributions which cannot easily be changed, mainly products stemming from the agricultural production.

The work energy exported as electricity as a consequence of the preceding need to be reduced by the 121 TJ y^{-1} invested in the substitution of fuel for cars, oil boilers and industrial consumption and additional 30 TJ y^{-1} for the ferry—resulting in an export 187 TJ y^{-1} (including the 5% increase in production).

Solid wastes are estimated to involve about 40 TJ y^{-1} of work energy in garbage from households and gardens and organics from industry, which, in this scenario, will be recycled on the island.

The work energies imported are now adding up to a total of 1,281 TJ y^{-1}. For energy or matter, the contributions are estimated to be 567 TJ y^{-1} (44%) and 714 TJ y^{-1} (56%) as energy- or matter-bound work energies, respectively. Out of this, 651 TJ y^{-1} (51%) are considered to originate in renewable resources, and 811 TJ y^{-1} (49%) are considered to be non-renewables.

The outputs sum to a total of 1,434 TJ y^{-1}. Out of this, 187 TJ y^{-1} (13%) represents the export of the excess energy produced from windmills. The remaining 1,247 y^{-1} (87%) are all material-bound work energy mainly in the exported goods (agricultural products) and solid wastes. From the total, 1,043 TJ y^{-1} (73%) are considered to be renewables and 391 TJ y^{-1} (27%) to be non-renewables.

The indicators for the hypothetical 2020 scenario come out with the following values:

Stock indicator = 15,560/1,281 = 12.0

Renewable/non-renewable ratio = 651/630 = 1.03

Renewability efficiency = 651/1,281 = 0.51

O/I indicator = 1,434/1,281 = 1.12

The stock indicator has risen from 11 to 12 (9%), which demonstrates a development towards increasing sustainability as the same infrastructure is now maintained by fewer inputs.

Both the renewable/non-renewable ratio and the renewability efficiency have increased from 0.73 to 1.03 and from 0.43 to 0.51, respectively, meaning that 51% of the work energies from all resources will be coming from renewable resources if the steps described earlier are taken.

At the same time, the O/I indicator has decreased slightly also, indicating an increase in sustainability. That this indicator is less sensitive is caused by the fact that the output is derived from the exported agricultural products of the island. This export is the key to socio-economical sustainability.

All in all, the islanders may take a first step towards being free of fossil fuels and thus towards sustainability by starting a conversion following the strategy sketched earlier. Meanwhile, the greatest obstacle to being fully sustainable is the challenges posed by the material flows of the society and the flows that are dependent on non-renewables in particular.

11.9 Sub-Conclusions

A set of indicators has been established that may be monitored and possibly used for decision making on what initiatives to implement, in what sequence and so on. The indicators may be used at various levels of details, but their behaviour is believed to be universal. In addition, the indicators allow for the establishment of scenarios which allow us or, rather, the politicians and managers to look out into future and decide what regulatory or institutional measures must be taken in order to achieve a sustainable society. The list of indicators is not exhaustive; only our minds set the limits.

12

Conclusions and Perspectives

12.1 Conclusions

This chapter attempts to summarize the overall achievements of the project, first of all, the development of a tool which may allow us to analyze, estimate, monitor and evaluate the sustainability of societies. Whereas the development of the methodology and the first implementation have taken entrance points in society on a relatively small scale—namely the Danish island of Samsø, with an area of 114.26 km² and population of 3,885 persons—the resulting framework is assumed to be applicable to societies at almost any scale. Now an attempt of upscaling is taking place in Sweden in the county of Jämtland. The area of Jämtland (49,000 km²) is 12% larger than Denmark (43,000 km²) but has a population of 115,000 inhabitants only as compared to 5.8 million.

The applicability to other societal systems is assumed since the overall scope has included considerations on the generality of the method. This has from time to time caused problems as it may be difficult to obtain factual knowledge about the existence of patterns which are eventual common to all societal systems. Hence, the partitioning of our societies in sectors has been based not on single references but, rather, on common knowledge and sense. Maybe, with the exception of the energy system, our knowledge of society and sustainability seems often to be based on fragmented of individual activities, such as transport, food production and so on, rather than creating an understanding of our society as a whole, for example based on the organization of material or substance fluxes (MFA/SFA) of the system.

In the following, I list the conclusions reached—general as well as specific—and include some suggestions or recommendations to the extent that they may serve to improve further implementations or expansions of the method.

The case chosen—to analyze the situation on the island of Samsø during 2011—represents an interesting study in the overall context of addressing the issue of "How to become sustainable?" because the island is using a relatively high percentage of renewable energy already. The same is valid for the project in Sweden, where all counties are now supposed to become sustainable based on their own findings' initiatives.

12.2 Method, Tool and Indicators

- In order to govern the transition of a society towards a sustainable state, it is valuable to have a tool that may assist in the evaluation of performance over time, for instance by observing some indicators of societal efficiency.
- The overall idea behind this project has been (1) to develop a method and tool which (2) focuses on pools and exchanges of work energies and (3) possesses a generality to such a degree that it is possible to transfer it with relative ease to most European countries or larger regions.
- To develop such a tool requires the construction of methodology that comprises a high degree of stringency and consistency in its approach—the concept of work energy provides such a background and serves to unify the two major concerns with supply-side sustainability (Allen et al., 2003), namely energy and materials.

12.3 Work Energy and Sustainability Strategy

- First of all, it is an obvious idea to base such a tool on measures of functionality in terms of work energy as work energy makes it possible to compare the respective importance of both energetic and material cycles—work energy puts the two on the same currency and allows us to weigh the two systems against each other.
- Furthermore, the potential future development of a societal economy which has its valuation system focusing on consumption—amounts of and rates of use and destruction—will serve to construct an economic system compatible with true environmental sustainability.
- Work energy focuses on amounts of energy which can deliver work— such energies should not be lost but used until fully exhausted.
- Focusing on work energy reveals where in our society—we invest and consume large amounts of work energy—and at high rates— or use work energy with a low efficiency. This together with

observations on whether the work energies are based on renewable or non-renewable resources will guide society in a direction towards a more sustainable situation. The earlier mentioned indicators should tell about performance and serve to direct development.

- Both considerations on the necessity of a certain use of a resource and considerations on its scarcity (or ease with which it is produced, reduced, reused, recovered or recycled) will enter the background of decision making in specific cases. In particular, the potentials available through technology and the economy will enter the discussion here.

- Putting all aspects together will help in giving priorities—the most efficient and feasible action should be taken first. This process may also for a time be used for determining where to put subsidies.

- The society is interested in work energy—not in the energy that cannot do work as, for instance, the heat lost to the environment. It is therefore crucial to map the flow of work energy that can be used to indicate where we can use the work energy better and lose less as heat into the environment.

12.4 Society as Sectors

- An analysis based on a fully aggregated society may give a picture that is too condensed, and it is therefore considered to be necessary to divide the society in a set of basic sectors and analyze their relative importance.

- The division of society made here has been confirmed as being representative of a similar distinction has been identified elsewhere in current literature. Basically, nature is observed as opposed to society.

- Many societies are already able to present a relatively detailed and elaborate energy budget—as it is the case also with the municipality of Samsø—which makes an excellent platform to start out with. Meanwhile, a conversion of the energy budget, together with a mapping of material flows, needs to be carried out in order to construct a full picture of the state in terms of work energies.

- Such an attempt would draw the attention not only to specific sectors but also to the activities undertaken and may eventually lead to even more detailed studies. An interpretation for the future should be realized at best within the framework of cleaner production and/ or industrial ecology.

12.5 Scale Dependency

- In this report, it is demonstrated how it is possible to develop such a methodology, considering the availability of data in most countries.

- Meanwhile, it has been considered too complex to start with the development of such a tool on a full society, for example a whole country, and consequently, a model society has been chosen.

- Islands are nice models—boundaries are clear and fixed, and it should be relatively easy to determine fluxes to and from the system.

- The method has been tested on the island of Samsø—which may be considered a minimum model to imitate a full society.

The experiences from the attempts to implement the method on a larger scale in Sweden clearly shows that some important activities very quickly become troublesome to analyze, for instance, how to evaluate the role of transports and traffic in general to regulate and shape our society, as well as where to account for the loading on sustainability.

12.6 Data Availability

- Data collection is a time-consuming process. The present project has contributed to the facilitation of the elaboration of a sustainability evaluation by (1) the method taking an entrance point in geographical data (2) by identifying major feasible routes for mapping work energy of the system both in terms of (3) by the identifying and setting forward additional data needed.

- Data availability *sensu lato* has been found not always to be optimal. This concerns not only the way data have been organized but also the way they may be extracted or even merged. Much information which could be organized and structured electronically had to be arranged by re-entering data or merging data manually. The new laws on data handling may put an end to such activities in the future.

- Facilitating data availability with the purpose of improving the possibility to establish a societal sustainability analysis should receive increased attention in the future. Open access to data for research should be obvious but is now heavily threatened not to say impossible due to the commercialization of governmental entities handling the data.

- Again, upscaling may reveal additional needs of data. In particular, the expansion to other countries with other crops will change the demand for specific agricultural data. Most societies at larger scales will also tend to comprise more types of industries wherefore a larger amount of information on materials (substances, elements) is needed than was necessary here. This is another argument for establishing a common database.

12.7 Observations—General and Specific

- The division of society in sectors is a sensible decision and facilitating the determination of the respective importance. The relative importance will vary among societies. In the case of Samsø, when looking at geographical space, agriculture becomes the dominating sector. In a county like Jämtland, Sweden, forestry suddenly takes over both as a nature-like activity and as a central player in the industry (paper production), as well as a supplier of building materials and biofuels. The enormous potential for carbon sequestration is ignored here.

- Infrastructure is important as the amount of work energies invested, and in fact, it may represent large stocks stored in society for a shorter or longer time. Not considering the turn-over time the stocks will need to be replaced for the continued existence of a society. It is therefore needed to spend (consume) work energy to maintain the infrastructure in a good shape.

- Using the sectorial approach taken here, some sectors are bound to come out which appear to be relatively inefficient in performance, such as, for instance, is the case with the societal sector, that is the social sustainability part. Meanwhile, this may reveal a point where material approaches fall short as some sectors simply serve as information stores and transformers. The point of information in society is still not captured by this methodology. Such studies should be improved in future.

- An increasing amount of specific knowledge in the area of combining work energy analysis to production processes is at present produced by the scientific society (Dincer, Rosen, Valero, Sciubba, Wall). An attempt to preserve and expand these studies with the method presented here in a more systematic way should be established for the continuation of this work. This could at best be done by institutionalizing efforts in the area—and work energy has through this study been demonstrated to be useful as a core concept in the evolution of our societies in the context of sustainability.

- The work energy approach seems to stress also the importance of waste, maybe merely because it integrates the material-bound work energies in the analysis. To Samsø, in particular, it seems that a relatively high amount of valuable resources is leaving the island in this manner.

It has been demonstrated that the work energy analysis can be used to assess quantitatively what in this case Samsø has obtained by shifting to the use of more renewable energy to ensure a more sustainable development. Furthermore, it has by a scenario been demonstrated that the island could take a big step towards the state of being independent of fossil fuels by simply exploiting the presently existing surplus of work energy from existing wind turbines. In both the case of Samsø and of Jämtland, a net export of electricity is realized—in Jämtland due to excessive hydropower production. In both cases, the excessive production could be invested in making the societies sustainable.

In general, if the work energy is predicted to increase by the realization of a project, the project will change the development towards more sustainability, and the project should be recommended. To the opposite, if the work energy is predicted to decrease (increasing consumption of work energy) by the implementation of the project, it will become more difficult to reach sustainable development, and it cannot be recommended from an environmental sustainability point of view to realize the project.

12.8 Suggested Implementation Scheme

The following are suggestions for how to use the methodology on implementation:

- Identify quantitatively important amounts of work energy—this goes for both inputs, stocks and outputs.
- Identify the sectors responsible for excessive use and evaluate how important they are to other sides of sustainability; that is there might be necessary costs.
- For inputs—evaluate the necessity and consider the fact that the imported amount may be reduced by relatively simple means. Consider to what extent the input is composed of renewable or non-renewable resources and to what extent is it possible to reduce the inputs by already-existing and preferably sustainable technologies.

- For stocks—consider composition and consider that everything, in principle, needs to be replaced over time. Evaluate potential turnover times and future development, in particular for stocks composed of or based on non-renewable resources.

- For outputs—consider amounts and their respective destinies. Are they exported and valued elsewhere or are they simply lost (dissipated)? Consider possibilities to retain the work energies within the system, to reduce outputs and to reuse or recover or recycle within the system, in the same manner as for the inputs.

- Prioritize actions by considering the feasibility and existing technological levels.

- Take action in a prioritized manner—and consider again several types of sustainability. Cheap actions which lead to a large reduction in losses are preferable. One cannot ignore economics, but many studies in cleaner production have demonstrated that initiatives which are beneficial to the environment often goes hand in hand with economic and social benefits.

- Observe the sustainability state of the society over time by repeating points and, for instance, estimate some of the suggested indicators for sustainability.

- Continue acting and adjusting, making the indicators move in the "right direction"—until the indicators reach a level indicating that full sustainability has been reached.

References

Abel, T. 1998. Complex adaptive systems, evolutionism, and ecology within anthropology: Interdisciplinary research for understanding cultural and ecological dynamics. *Georgia Journal of Ecological Anthropology*, 2, 6–29.

Allen, T.F.H., Tainter, J.A. and Hoekstra, T.W. 2003. *Supply-Side Sustainability*. Columbia University Press, New York, 459 pages.

Ancker, M.-L., Fogh, A., Hansen, O.K., Hestbech, M., Justesen, P., Martin, H.L., Martinussen, H., Møller, J., Nielsen, V.F., Spleth, P., Thøgersen, R. and Aaes, O. 2011. *Håndbog i kvæghold*. Videnscentret for Landbrug, Landbrugsforlaget, 204 pages.

Aoki, I. 2018. Entropy principle for the evolution of living systems and the universe—from bacteria to the universe. *Journal of the Physical Society of Japan*, 87, 104801.

Apaiah, R.K., Linnemann, A.R. and Kooi, H.J. van der. 2006. Exergy analysis: A tool to study the sustainability of food supply chains. *Food Research International*, 39, 1–11.

Ayres, R.U. 1998. Eco-thermodynamics: Economics and the second law. *Ecological Economics*, 26, 189–209.

Ayres, R.U. 1999. The second law, the fourth law, recycling and limits to growth. *Ecological Economics*, 29, 473–483.

Baccini, P. and Brunner, P.H. 2012. *Metabolism of the Anthroposphere: Analysis, Evaluation, Design*. The MIT Press, Cambridge, MA, 392 pages.

Barbier, E.B. 2019. The concept of natural capital. *Oxford Review of Economic Policy*, 35, 1, 14–36.

Barbier, E.B. and Burgess, J.C. 2017. Natural resource economics, planetary boundaries and strong sustainability. *Sustainability*, 9, 1858.

Bastianoni, S. 1998. A definition of 'pollution' based on thermodynamic goal functions. *Ecological Modelling*, 113, 163–166.

Bastianoni, S. and Marchettini, N. 1997. Energy/exergy ratio as a measure of the level of organization of systems. *Ecological Modelling*, 99, 33–40.

Bastianoni, S., Nielsen, S.N., Marchettini, N. and Jørgensen, S.E. 2005. Use of thermodynamic functions for expressing some relevant aspects of sustainability. *International Journal of Energy Research*, 29, 53–64.

Bastianoni, S., Pulselli, F.M. and Rustici, M. 2006. Exergy versus energy flow in ecosystems: Is there an order in maximizations? *Ecological Indicators*, 6, 1, 58–62.

Bauman, H. and Tillman, A.-M. 2004. The Hitch Hiker's guide to LCA. *Studentlitteratur*, 543.

Bendoricchio, G. and Jørgensen, S.E. 1997. Exergy as goal function of ecosystems dynamic. *Ecological Modelling*, 102, 1, 5–15.

Brillouin, L. 1962. *Science and Information Theory* (2nd edition of work published 1956). Dover Publications, New York, 347 pages.

Campbell, C.J. 1997. *The Coming Oil Crisis*. Multi-Science Publishing.

Carnot, S. 1824. *Reflections on the Motive Power of Heat* (translated 1943). ASME, New York, 107 pages.

Carson, R. 1962. *Silent Spring*. Houghton Mifflin Company.

Chen, G.Q. and Chen, B. 2009. Extended-exergy analysis of the Chinese society. *Energy*, 34, 1127–1144.

Clausius, R. 1865. Über verschiedene für die Anwendung bequeme Formen der Hauptgleichungen der mechanischen Wärmetheorie. *Annalen der Physik*, 125, 7, 353–400.

Cleveland, C. and Ruth, M. 1997. *When, Where, and by How Much Do Biophysical Limits Constrain the Economic Process?* A survey of Nicholas Georgescu-Roegen's contribution to ecological economics.

Cornelissen, R.L. and Hirs, Gerard G. 2002. The value of the exergetic life cycle assessment besides the LCA. *Energy Conversion and Management*, 43, 9–12, 1417–1424.

Dalgaard, T., Dalgaard, R. and Højlund Nielsen, A. 2002. *Energiforbrug på økologiske og konventionelle landbrug.* Afdeling for Jordbrugssystemer, Ministeriet for Fødevarer, Landbrug og Fiskeri. Danmarks JordbrugsForskning, Markbrug 269, 8 pages.

Daly, H.E. 1992. Is the entropy law relevant to the economics of natural resource scarcity? Yes, of course it is! *Journal of Environmental Economics and Management*, 23, 91–95.

Daly, H.E. 1995. On Nicholas Georgescu-Roegen's contributions to economics: An obituary essay. *Ecological Economics*, 13, 149–154.

Daly, H.E. and Cobb, J.B. 1989. *For the Common Good: Redirecting the Economy Toward Community, the Environment, and a Sustainable Future.* Beacon Press, Boston, 1994, 534 pages.

Danmarks Statistik. 2011. *Statistisk Årbog, 2011.* Danmarks Statistik, Copenhagen, 525 pages.

Deacon, T.W. 2007. Shannon-Boltzmann-Darwin: Redefining information. Part 1. *Cognitive Semiotics*, 1, 123–148.

Deacon, T.W. 2008. Shannon-Boltzmann-Darwin: Redefining information. Part 2. *Cognitive Semiotics*, 2, 167–194.

Dewulf, J., Langenhoeve, H. van, Muys, B., Bruers, S., Bakshi, B.R., Grubb, G.F., Paulus, D.M. and Sciubba, E. 2008. Exergy: Its potential and limitations in environmental science and technology. *Environmental Science & Technology*, 42, 7, 2221–2232.

Dincer, I. 2002. The role of exergy in policy making. *Energy Policy*, 30, 137–149.

Dincer, I. and Rosen, M.A. 2007. Energetic, exergetic, environmental and sustainability aspects of thermal energy storage systems. In *Thermal Energy Storage for Sustainable Energy Consumption: Fundamentals, Case Studies and Design* (Ed.: H.Ö. Paksoy). NATO Science Series book series, NAII, volume 234, pp. 23–46.

EEA. 2007. *CLC2006 Technical Guidelines.* EEA Technical Report, No. 17, 70 pages.

EEA. 2013. *Late Lessons from Early Warnings: Science, Precaution, Innovation.* Summary. EEA Report, No. 1, 48 pages.

Ertesvåg, I.S. 2001. Society exergy analysis: A comparison of different societies. *Energy*, 26, 253–270.

Ertesvåg, I.S. 2005. Energy, exergy, and extended-exergy analysis of the Norwegian society. *Energy*, 30, 649–675.

Finnveden, G., Johansson, J., Lind, P. and Moberg, Å. 2005. Life cycle assessment of energy from solid waste—part 1: General methodology and results. *Journal of Cleaner Production*, 13, 213–229.

Fonseca, J.C., Marques, J.C., Paiva, A.A., Freitas, A.M., Madeira, V.M.C. and Jørgensen, S.E. 2000. Nuclear DNA in the determination of weighting factors to estimate exergy from organism's biomass. *Ecological Modelling*, 126, 179–190.

Gaudreau, K., 2009. *Exergy Analysis and Resource Accounting. Master Thesis, Master of Environmental Studies in Environment and Resource Studies*. Waterloo, Ontario, Canada, 128 pages.

Georgescu-Roegen, N. 1971. *The Entropy Law and the Economic Process*. Harvard University Press, Cambridge.

Glansdorff, P. and Prigogine, I. 1971. *Thermodynamic Theory of Structure, Stability and Fluctuations*. Wiley-Interscience, New York.

Gong, M. and Wall, G. 2001. On exergy and sustainable development—part 2: Indicators and methods. *Exergy*, 1, 4, 217–233.

Gong, M. and Wall, G. 2016. Exergy analysis of the supply of energy and material resources in the Swedish society. *Energies*, 9, 707, doi:10.3390/en90900707

Gowdy, J. and Mesner, S. 1998. The evolution of Georgescu-Roegen's bioeconomics. *Review of Social Economy*, LVI, 2, 136–156.

Habermas, J. 1992. *The Structural Transformation of the Public Sphere: Inquiry into a Category of Bourgeois Society*. New Edition. Polity Press, 328 pages.

Hammond, G.P. 2004. Engineering sustainability: Thermodynamics, energy systems, and the environment. *International Journal of Energy Research*, 28, 613–639.

Hammond, G.P. 2007. Industrial energy analysis, thermodynamics and sustainability. *Applied Energy*, 84, 675–700.

Hannon, B., Ruth, M. and Delucia, E. 1993. A physical view of sustainability. *Ecological Economics*, 8, 253–268.

HDP, 2008. *Håndbog i driftsplanlægning 2008 (Manual of Operations Planning)*. Landbrugsforlaget, Århus N, 200 pages.

Hellström, D. 1997. An exergy analysis for a wastewater treatment plant- an estimation of the consumption of resources. *Water Environment Research*, 69, 44ff.

Hirs, G. 2003. Thermodynamics applied. Where? Why? *Energy*, 28, 1303–1313.

Holland, J.H. 1992. Complex adaptive systems. *Daedalus*, 121, 1, 17–30.

Hovelius, K. 1997. *Energy-, Exergy- and Emergy Analysis of Biomass Production*. Institutionen for lantbruksteknik, Rapport 222. Swedish University of Agricultural Sciences, Uppsala.

Hubbert, M.K. 1962. *Energy Resources*. National Academy of Sciences, Publication 1000-D, 60p.

Ignatenko, O., van Schaik, A. and Reuter, M.A. 2007. Exergy as a tool for evaluation of the resource efficiency of recycling systems. *Minerals Engineering*, 20, 9, 862–874.

Jaynes, E.T. 1957a. Information theory and statistical mechanics. *Physical Review*, 106, 4, 620–630.

Jaynes, E.T. 1957b. Information theory and statistical mechanics. II. *Physical Review*, 108, 2, 171–190.

Jonas, H. 1984. *The Imperative of Responsibility: In Search of Ethics for the Technological Age*. (German 1979), University of Chicago Press, 263 pages.

Jørgensen, K. (Ed.) 2008. *Håndbog i driftsplanlægning 2008*. Dansk Landbrugsrådgivning, Landscentret, Landbrugsforlaget, 200 pages.

Jørgensen, S.E. 1992. *Integration of Ecosystem Theories: A Pattern*. Springer.

Jørgensen, S.E. 2006. *Eco-Exergy as Sustainability*. WITPRESS, Southampton, 207 pages.

Jørgensen, S.E. 2012. *Integration of Ecosystem Theories: A Pattern*. Kluwer, Dordrecht, 388 pages.

Jørgensen, S.E., Ladegaard, N., Debeljak, M. and Marques, J.C. 2005. Calculation of exergy for organisms. *Ecological Modelling*, 185, 165–175.

Jørgensen, S.E., Ludovisi, A. and Nielsen, S.N. 2010. The free energy and information embodied in the amino acid chains of organisms. *Ecological Modelling*, 221, 19, 2388–2392.

Jørgensen, S.E. and Mejer, H.F. 1977. Ecological buffer capacity. *Ecological Modelling*, 3, 39–61.

Jørgensen, S.E. and Mejer, H.F. 1979. Holistic approach to ecological modelling. *Ecological Modelling*, 7, 169–189.

Jørgensen, S.E. and Mejer, H.F. 1981. Application of exergy in ecological models. In *Progress in Ecological Modelling* (Ed.: D. Dubois), Elsevier, Amsterdam, pp. 39–47.

Jørgensen, S.E. and Nielsen, S.N. 2015. A carbon cycling model developed for the renewable Energy Danish Island, Samsø. *Ecological Modelling*, 306, 106–120.

Jørgensen, S.E., Nielsen, S.N. and Mejer, H. 1995. Emergy, environ, exergy and ecological modelling. *Ecological Modelling*, 77, 99–109.

Jørgensen, S.E. and Svirezhev, Y.M. 2004. *Towards a Thermodynamic Theory for Ecological Systems*. Elsevier, Amsterdam, 366 pages.

Karnani, M. and Annila, A. 2009. Gaia again. *BioSystems*, 95, 82–87.

Kaysen, O. and Petersen, C. (Econet AS). 2010. *Vurdering af genanvendelsesmålsætninger i affaldsdirektivet*. Miljøprojekt, Nr. 1328. Miljøministeriet, Miljøstyrelsen, 107 pages.

Koroneos, C. and Kalemakis, I. 2012. Exergy indicators in the building environment. *International Journal of Exergy*, 11, 4, 439–459.

Koroneos, C., Roumbas, G. and Moussiopoulos, N. 2005. Exergy analysis of cement production. *International Journal of Exergy*, 2, 1, 55–68.

Koroneos, C., Spachos, T. and Moussiopoulos, N. 2003. Exergy analysis of renewable energy sources. *Renewable Energy*, 28, 295–310.

Koroneos, C.J., Nanaki, E.A. and Xydis, G.A. 2012. Sustainability indicators for the use of resources—the exergy approach. *Sustainability*, 4, 1867–1878.

Kriebel, D., Tickner, J., Epstein, P., Lemons, J., Levins, R., Loechler, E.L., Quinn, M., Rudel, R., Schettler, T. and Stoto, M. 2001. The precautionary principle in environmental science. *Environmental Health Perspectives*, 109, 9, 871–876.

Kullback, S. and Leibler, R.A. 1951. On information and sufficiency. *The Annals of Mathematical Statistics*, 22, 1, 79–86.

Lotka, A.J. 1922a. Contribution to the energetics of evolution. *Proceedings of the National Academy of Sciences*, 8, 147–150.

Lotka, A.J. 1922b. Natural selection as a physical principle. *Proceedings of the National Academy of Sciences*, 8, 151–154.

Lovelock, J. 1979. *Gaia: A New Look at Life on Earth*. Oxford University Press, Oxford, 176 pages.

Ludovisi, A. 2009. Exergy vs. information in ecological successions: Interpreting community changes by classical thermodynamic approach. *Ecological Modelling*, 220, 1566–1577.

Ludovisi, A. and Jørgensen, S.E. 2009. Comparison of exergy found by a classical thermodynamic approach and by the use of the information stored in the genome. *Ecological Modelling*, 220, 1897–1903.

Ludovisi, A., Pandolfi, P. and Taticchi, M.I. 2005. The strategy of ecosystem development: Specific dissipation as an indicator of ecosystem maturity. *Journal of Theoretical Biology*, 235, 33–43.

McMahon, G.F. and Mrozek, J. 1997. Economics, entropy and sustainability. *Hydrological Sciences Journal/Journal des Sciences Hydrologiques*, 42, 4, 501–512.

Meadows, D., Meadows, D.L., Randers, J. and Behrens III, W.W. 1972. *The Limits to Growth*. Potomac Associates, Earth Island limited, London, 205 pages.

Meester, B. de, Dewulf, J., Verbeke, S., Janssens, A. and Van Langenhove, H. 2009. Exergetic life-cycle assessment (ELCA) for resource consumption evaluation in the built environment. *Building and Environment*, 44, 11–17.

Mejer, H.F. and Jørgensen, S.E. 1979. Exergy and ecological buffer capacity. In *State of the Art in Ecological Modelling* (Ed.: S.E. Jørgensen). ISEM, Copenhagen and Pergamon Press, Oxford, pp. 829–846.

Mester, B. de, Dewulf, J., Janssens, A. and Langenhoeve, H. van, 2006. An improved calculation of the exergy of natural resources for exergetic life cycle assessment (ELCA). *Environmental Science & Technology*, 40, 6844–6851.

Moberg, Å., Finnveden, G., Johansson, J. and Lind, P. 2005. Life cycle assessment of energy from solid waste—part 2: Landfilling compared to other treatment methods. *Journal of Cleaner Production*, 13, 231–240.

Mora, C.H. and Oliveira Jr., S. de. 2006. Environmental exergy analysis of wastewater and treatment plants. *Engenharia Térmica*, 5, 2, 24–29.

Nicolis, G. and Prigogine, I. 1977. *Self Organization in Non-Equilibrium Systems*. Wiley-Interscience, New York, 491 pages.

Nielsen, A. (Carl Bro A/S) Statens Byggeforskningsinstitut, 1993. *Byggeriets materialeforbrug. Registrering af bygningsdele og byggematerialer, som hare en særlig betydning for indførelsen af renere teknologi*. Miljøprojekt, nr. 221. Miljøministeriet, Miljøstyrelsen, 96 pages.

Nielsen, S.N. 2007. What has modern ecosystem theory to offer cleaner production, industrial ecology or the society? The views of an ecologist. *Journal of Cleaner Production*, 15, 17, 1639–1653.

Nielsen, S.N. and Bastianoni, S. 2007. A common framework for emergy and exergy based LCA in accordance with environ theory. *International Journal of Ecodynamics*, 2, 3, 170–185.

Nielsen, S.N. and Jørgensen, S.E. 2013. Goal functions, orientors and indicators (GoFOrIt's) in ecology: Application and functional aspects—strengths and weaknesses. *Ecological Indicators*, 28, 31–47.

Nielsen, S.N. and Müller, F. 2009. Understanding the functional principles of nature—proposing another type of ecosystem services. *Ecological Modelling*, 220, 1913–1925.

Nielsen, S.N., Fath, B., Bastianoni, S., Marques, J.C., Muller, F.M., Patten, B.C., Ulanowicz, R.E., Jørgensen, S.E. and Tiezzi, E. 2020. *A New Ecology (2nd ed.)*. Elsevier, Amsterdam, 270 pages.

Odum, E.P. 1953. *Fundamentals of Ecology*. W. B. Saunders Company, Philadelphia, 383 pages.

Odum, E.P. 1969. The strategy of ecosystem development. *Science*, 164, 262–270.

Pedersen, L. (ed.), 2011. *Håndbog i kvæghold (Manual for livestock)*. Landbrugsforlaget, Århus N, 204 pages.

Pimentel, D. (Ed.) 1980. *Handbook of Energy Utilization in Agriculture*. CRC Press, Boca Raton, 475 pages.

Pimentel, D. and Pimentel, M. (Eds.) 1996. *Food, Energy, and Society*. Revised edition (1979). University Press of Colorado, Niwot.

Prigogine, I. and Stengers, I. 1984. *Order Out of Chaos: Man's New Dialogue with Nature*. Fontana Paperbacks, Flamingo, 349 pages.

Rasmussen, L.B. and Sylvestersen, H.L., (eds.), 2009. *Håndbog i plantedyrkning (Manual of plant growing)*. Landbrugsforlaget, Århus N, 200 pages.

Rechberger, H. and Graedel, T.E. 2002. The contemporary European copper cycle: Statistical entropy analysis. *Ecological Economics*, 42, 59–72.

Rifkin, J. 1989. *Entropy: Into the Greenhouse World*. Revised edition. Bantam Books, New York, 354 pages.

Rivero, R. and Garfias, M. 2006. Standard chemical exergy of elements updated. *Energy*, 31, 3310–3326.

Ruth, M. 1995. Information, order and knowledge in economic and ecological systems: Implications for material and energy use. *Ecological Economics*, 13, 99–114.

Schrödinger, E. 1944. *What Is Life? The Physical Aspect of the Living Cell*. Cambridge University Press, Cambridge, 194 pages.

Sciubba, E. 2003. Extended exergy accounting applied to energy recovery from waste: The concept of total recycling. *Energy*, 28, 1315–1334.

Sciubba, E. 2004. Exergoeconomics. *Encyclopedia of Energy*, 2, 577–591. Elsevier.

Sciubba, E. and Wall, G. 2007. A brief commented history of exergy: From the beginnings to 2004. *International Journal of Thermodynamics*, 10, 1, 1–26.

Serova, E.N. and Brodiansky, V.M. 2004. The concept "environment" in exergy analysis: Some special cases. *Energy*, 29, 2397–2401.

Skene, K.R. 2015. Life's a gas: A thermodynamic theory of biological evolution. *Entropy*, 17, 8, 5522–5548.

Skytt, T., Nielsen, S.N. and Fröling, M. 2019. Energy flows and efficiencies as indicators of regional sustainability—a case study of Jämtland, Sweden. *Ecological Indicators*, 100, 74–98.

Skytt, T., Nielsen, S.N. and Jonsson, B.-G. 2020. Global warming potential and absolute global temperature change potential from carbon dioxide and methane fluxes as indicators of regional sustainability – A case study of Jämtland, Sweden. *Ecological Indicators*, 110, 105831.

Stjernholm, M. 2009. *Corine Land Cover 2006*. Final report on interpretation of CLC2006 in Denmark. Research Notes from NERI No. 257. National Environmental Research Institute, Aarhus University, Denmark, 48 pages. Research Notes from NERI No. 257. www.dmu.dk/Pub/AR257.pdf.

Svensson, N., Roth, L., Eklund, M. and Mårtensson, A. 2006. Environmental relevance and use of energy indicators in environmental management and research. *Journal of Cleaner Production*, 14, 134–145.

Swenson, R. 1989. Emergent evolution and the global attractor: The evolutionary epistemology of entropy production maximization. *Systems Sciences*, 33, 3, 46–53.

Szargut, J. 2005. *Exergy Method: Technical and Ecological Applications*. WIT Press, Southampton, 164 pages.

Szargut, J., Morris, D.R. and Steward, F.R. 1988. *Exergy Analysis of Thermal, Chemical and Metallurgical Processes*. Hemisphere Publishing Corporation, New York.

UNESCO/UNCED. 1992. *The Rio Declaration on Environment and Development*. www.unesco.org/education/pdf/RIO_E.PDF.

Valderrama, C., Granados, R., Cortina, J.L., Gasol, C.M., Guillem, M. and Josa, A. 2013. Comparative LCA of sewage sludge valorisation as both fuel and raw material substitute in clinker production. *Journal of Cleaner Production*, 51, 205–213.

Valero, A., Usón, S., Torres, C. and Valero, A. 2010. Application of thermoeconomics to industrial ecology. *Entropy*, 12, 591–612.

Valero, A., Valero, A. and Martínez, A. 2010. Inventory of the exergy resources on earth including its mineral capital. *Energy*, 35, 989–995.

Vium, J. 2006. *Affaldsforebyggelse ved renovering.* Arbejdsrapport fra Miljøstyrensen, Nr. 42. Miljøministeriet, Miljøstyrelsen, 79 pages.

Volk, T. and Pauluis, O. 2010. It is not the entropy you produce, rather how you produce it. *Philosophical Transactions of the Royal Society B*, 365, 1317–1322.

Vosough, A., Noghrehabadi, A., Ghalambaz, M. and Vosough, S. 2011. Exergy concept and its characteristic. *International Journal of Multidisciplinary Sciences and Engineering*, 2, 47–52.

Wall, G. 1977. *Exergy—a Useful Concept Within Resource Accounting.* Report No. 77-42. Institute of Theoretical Physics, Chalmers University of Technology and University of Göteborg, Sweden, 61 pages.

Wall, G. and Gong, M. 2001. On exergy and sustainable development—part 1: Conditions and concepts. *Exergy*, 1, 3, 128–145.

WCED. 1987. *Our Common Future.* Brundtland Report. UN, 374 pages.

Whittaker, R.H. 1975. *Communities and Ecosystems.* 2nd edition. Macmillan Publishing Co., New York, xvii + 385 pages.

Zhou, C., Hu, D., Wang, R. and Liu, J. 2011. Exergetic assessment of municipal solid waste management system. *Ecological Complexity*, 8, 171–176.

Index

Printed and bound by CPI Group (UK) Ltd, Croydon, CR0 4YY

23/10/2024

01778223-0014